The Art of Cyberwar

The Principles of Conflict in Cyberspace

Michael A. VanPutte, Ph.D.

Lieutenant Colonel, U.S. Army (retired)

Thomas P. Sammel

Major, U.S. Marine Corps (retired)

Other books by Michael VanPutte

Walking Wounded – Inside the US Cyberwar Machine

We appreciate your suggestions, comments, and criticisms. You can reach us at michael@mvanputte.com or www.mvanputte.com.

Cover by Brianne VanTuyle

Copyright © 2019 Michael A. VanPutte and Thomas P. Sammel
All rights reserved.
ISBN-13: 978-1081107574

Dedications

Michael A. VanPutte

To the three women who have graciously put up with my idiosyncrasies: Linda, Ashley and Brianne.

Thomas P. Sammel

I dedicate this book to my incredible wife Linda, my three amazing children, Anna, Sabrina, and Andrew, my father Richard, and my late mother Mary. Thank you for everything you have done for me.

Contents

CHAPTER 1. INTRODUCTION ... 1
   PRINCIPLES OF WAR ... 2
CHAPTER 2. CYBERSPACE ... 5
   THE POWER OF INFORMATION ... 5
   UNINTENDED CONSEQUENCES ... 6
   VULNERABILITIES ... 8
   SPACE AND TIME ... 13
   SECURITY CONTROLS ... 15
   DEPENDENCIES ... 18
   PSYCHOLOGICAL EFFECTS ... 19
   DECISION MAKING ... 21
   PRIVACY AND ANONYMITY ... 23
   TRUST AND DISTRUST ... 27
   LEGAL AND CULTURAL COMPLEXITY ... 28
   VULNERABILITY AVAILABILITY ... 29
   WEAPON AVAILABILITY ... 31
   VULNERABILITY AND WEAPON LIFESPAN ... 33
   THE HUMAN VULNERABILITY ... 35
   RESILIENCE ... 36
CHAPTER 3. PRINCIPLES OF CONFLICT IN CYBERSPACE ... 39
   ANONYMITY ... 40
   AUDACITY ... 41
   INITIATIVE ... 43
   ASYMMETRIC EFFECTS ... 45
   ECONOMY OF FORCE ... 48
   RESILIENCE ... 51
   HETEROGENEITY ... 52
   DISPERSION OF EFFORT ... 53
CHAPTER 4. CONCLUSION ... 55
GLOSSARY ... 59
REFERENCES AND NOTES ... 63
INDEX ... 71
ABOUT THE AUTHORS ... 73

List of Illustrations

Figure 1 – An *Attacker's* Window of Vulnerability   33
Figure 2 – A Victim's Window of Vulnerability   34

Preface

"Those who cannot remember the past are condemned to repeat it."
— George Santayana (1863-1952)

Cyberspace has created profound changes in public and private organizations. As a result, many organizations are struggling to understand how to operate in an environment that enables nations, criminals, and hobbyists to exploit their reliance on this digital environment.

This book proposes to illustrate the properties of cyberspace and fundamental principles for the conflict that exists in cyberspace. This understanding will allow organizations to use cyberspace in a coherent and thoughtful way. This understanding will also allow security researchers to test and refine theories, optimize strategies and tactics, and evaluate new technologies and operations to improve their security in this ubiquitous domain.

Acknowledgements

We would like to express our gratitude to the many people who helped research and review this project. We would especially like to thank Mr. Richard Guidorizzi for his support throughout this effort.

Chapter 1. Introduction

Nam et ipsa scientia potestas est.
"Knowledge itself is power."
- Sir Francis Bacon (1561-1626)

Cyberspace has evolved from a research project used by government and academic scholars into a worldwide information environment critical to our national and international well-being. Public and private organizations have obtained enormous productivity and profitability gains by leveraging new and existing technologies and transitioning their processes into cyberspace. Individuals have benefited through vast online marketplaces, collaboration, and social media. World leaders have stepped up to promise the world would contract as cyberspace connected the world's populations into a global village.

While cyberspace has presented unparalleled opportunities to improve our way of life, a perpetual conflict is occurring and increasing with the potential to negate many of the benefits. The instruments of this maleficence are easily accessible and inexpensive. Public and private hackers need a few hundred dollars of hardware to cause mayhem. They can learn the prerequisite skills at any university or on the Internet itself. Evolving and effective attack tools are readily available online. At no time in the history of mankind has surveillance and destructive capabilities become so prolific and available to the public.

A better understanding of cyberspace, the principles that apply to cyberspace and the conflict that exists there may help organizations understand and address the challenges. As we go forward note that there are no universally accepted terms for the activities related to computer security, information assurance,

Introduction

computer network operations, cyber warfare, cyber terrorism, digital crime, and recreational hacking. Therefore, the authors will use the phrase cyber conflict to cover all of these activities.

Principles of War

After studying the history of war, military leaders have developed Principles of War as fundamental propositions guiding strategic warfare. These principles apply to *physical* armed conflict between rival groups, often referred to as *kinetic warfare*.

The U.S. military based their Principles of War on classical military strategy and theory, primarily Carl von Clausewitz's *On War*. These principles have been the foundation of U.S. kinetic warfighting for generations, and are:

- Objective – Direct every military operation toward a clearly defined, decisive, and attainable goal.
- Offense – Seize, retain, and exploit the initiative.
- Mass – Concentrate the effects of combat power at a decisive place and time.
- Economy of Force – Allocate minimal essential combat power to secondary efforts.
- Maneuver – Place the enemy in a disadvantageous position through the flexible application of combat power.
- Unity of Command – For every objective ensure unity of effort under one responsible commander.
- Security – Never permit the enemy to acquire an unexpected advantage.
- Surprise – Strike the enemy at a time, place, or in a manner for which they are unprepared.
- Simplicity – Prepare clear, uncomplicated plans and clear, concise orders to ensure a thorough understanding.

US Joint Publication 3-0 supplements these nine principles with three Principles of Joint Operations:

- Perseverance – Ensure the commitment necessary to attain the national strategic end state.
- Legitimacy – Develop and maintain the will necessary to attain the national strategic end state.

- Restraint – Limit collateral damage and prevent the unnecessary use of force.

The US military notes that their Principles of War are neither absolutes nor checklists for success. They are fundamental characteristics of successful military operations. Military leaders must weigh each principle against the situation they find themselves, and balance the principles with the realities of the conflict they face.

However, the Principles of War vary across nations based on the nation's experiences in warfare and their military culture. The Russian Principles of Military Science include *obedience*, dictating Russian leaders and soldiers follow orders without question. US military doctrine grants military leaders latitude in the means to pursue military objectives.

Numerous researchers have applied various principles to military cyber operations.[1] These studies have a serious shortcoming. Warfare involves nations or ideological organizations applying violence, or the threat of violence, in pursuit of their respective state or non-state goals in a geographic battlespace. Cyber conflict involves more than military operations.[2]

> Warfare might be closer to a social science than a physical science.

While military organizations conduct offensive and defensive cyber operations, intelligence organizations simultaneously seek to collect intelligence and perform covert operations to influence political, military, and economic conditions abroad.[3] Other organizations and individuals also perform operations against many of the same information systems; law enforcement agencies conduct domestic surveillance, counter-intelligence agencies monitor for spies; and counter-terrorism agencies search for terrorists and their support networks. Further, criminals steal personal, corporate, and government information for profit. Amateurs and hobbyists collect information, modify information and systems, and deny and degrade systems in the pursuit of personal or organizational goals. All of these entities use the same technology and tactics, on the same battlefield, at the same time.

Introduction

And there may be many instantiations of each of these entities as multiple nations, organizations, and individuals pursue their independent goals.

This complexity led to the investigation of these goals, and the creation of this book: to discover the fundamental principles of cyber-conflict regardless of the intent of the person or group conducting the operations.

Our study of the topic is an attempt to explain cyberspace and cyber-conflict to the layman. It does not advocate for or against government operations in cyberspace, or whether one government agency or department should or should not establish or act on national cyber priorities.

There is a wide variety of understanding and misunderstanding about cyberspace. So, before we discuss principles of cyberconflict we need to understand the characteristics of cyberspace.

Chapter 2. Cyberspace

This chapter describes the fundamental truths of cyberspace that enable and constrain cyber operations.

The Power of Information

Since the dawn of time, possessing the right information at the right time has enabled people to make informed decisions, and increase their ability to accomplish their goals.

Computing technologies enable people to access, store, and manipulate vast amounts of information. These technologies allow organizations to take immediate actions that a generation ago would have required phone calls or face-to-face transactions. Computers may also control mechanical devices. Computing technologies therefore have the potential to generate and amplify power by enabling people to make more informed and rapid decisions, or cause mechanical activities to occur without human intervention.

Researchers developed telecommunications technologies to transfer large amounts of information between computers. This spawned the Department of Defense to establish ARPANet. The follow-on Internet stemmed from the realization that internetworking had a distinct place in the world beyond governments and academia.

The merging of computing and telecommunication technologies transformed *computers into information concentrators and telecommunications media into information conductors.*

Our demand for the newest and fastest technologies has been insatiable. Moore's Law, established by the co-founder of Intel

Corporation Gordon Moore, states computing power expressed as the number of transistors in an integrated circuit, doubles approximately every two years.[1] Kryder's Law, created by the senior vice president of research and chief technology officer of the digital storage company Seagate, Mark Kryder, states hard drive storage density doubles every thirteen months.[2] Nielsen's Law created by web usability consultant Jakob Nielsen states a high-end user's home network connection speed doubles every 21 months. These three laws demonstrate how computing power, storage, and telecommunications capabilities are increasing at tremendous rates.

Computer scientist Robert Metcalf, co-inventor of the Ethernet communications standard proposed the value of an information network is proportional to the square of the number of connected users in the system.[3] As we add more individuals and systems to a network, each user on the network receives increased benefits, encouraging adding even more information and people. In December 2017 over 4.1 billion people had access to the Internet, representing over 54% of the world population.[4]

Consequently, cyberspace is the result of the cumulative efforts by governments, academia, businesses and individuals around the world. Each of these entities take part in and lay claim to the operations, maintenance, and ownership of portions of this environment. Each also has a vested interest in the conflict occurring within cyberspace.

Unintended Consequences

The rapid growth and ubiquity of computing platforms and the increased benefit from being 'connected' has created an information environment with a vast number of

> Information enables power; and access, use and reliance on information provides the impetus for conflict.

individuals and groups connected to each other. Inevitably some of these individuals and groups perceive their political, financial, or personal goals take greater precedent than other Internet residents. The result is a persistent conflict that has both winners and losers, as well as predators and victims.

People learned that being connected to a network results in having a connection with everyone on the network, and to their systems and information. As we connect larger numbers of diverse political, social, economic and cultural groups worldwide, we increase the potential for nefarious undertakings.

So, while computers and networking may improve lives, they also introduce new *threats*. The early days of the Internet tempted spies as well as curious hobbyists to access the military, academic, and personal information existing there.[5] Once commerce and banking moved into cyberspace, bringing with them the need to access vast repositories of personal and corporate wealth, bank robber Willie Sutton's words to "Go where the money is… and go there often" rang true. In 2017 alone, the FBI Internet Crime Complaint Center (IC3) received 301,580 complaints related to cybercrimes with a dollar loss exceeding $1.4 billion.[6] Taking this to a broader global audience, the 2017 Norton Cyber Security Insights Report Global Results states that cybercrime affected over 900 million people worldwide with monetary losses of $172 billion.[7]

Cyberspace enables organizations and individuals to pursue their goals and challenge their perceived adversaries. These battles occur in many forms including denial of service attacks, espionage, intellectual property theft, financial fraud, network breaches, destructive attacks, and deceptive social media posts. They may occur over a narrow time frame or across many years, and be engaged in by thrill seekers, hacktivists, criminals, ideologues, nation states, and others.

Military strategists have borrowed wholesale from existing history and literature regarding "kinetic" operations and attempted to cast cyberspace into their traditional mold as a 'domain' of warfare.[8,9] However, cyberspace is a man-made environment with significant differences from naturally created domains. These differences enabled a constant state of contestation in nearly every country around the world.

Vulnerabilities

The vast majority of computing systems purchased and deployed today are commercial, general-purpose systems. Profit motivates the vendors of these systems to deliver products that meet or create consumer demand. Many public and private consumers demand the newest, least expensive, and easiest to use computers and software. Commercial computer manufacturers and software developers have responded to this market demand by quickly building computers and the software to operates these computers as rapidly as possible to gain and maintain market share.

Consumers also demand that their systems are *backwards compatible*. This refers to the ability of new computers to run legacy software from old computers, and access legacy data files. The result is equivalent to taking old programs and bolting on new functions so legacy programs or data will continue to operate and be accessible. As we will see later, this creates serious security issues.

Consumers also want to modify their computers by adding new hardware and programs This *extensibility* requires computer companies to change their own computer code, which changes how the operating system and applications function. Extensibility also allows software developers to 'push' the latest "update" to computers over networks. This capability is written into the operating systems running computers, and the programs we use.

When a user installs a new printer or mouse, the computer may not have the instructions to operate the new device. A modern computer will install the device's *"drivers"* and applications to enable the device to communicate with the computer's operating systems. Likewise, when a user wants to use new multi-media online, their Internet browser may automatically search for, find, and install the latest *"plug-in"* software to enable them to access the new media efficiently.

The result of this demand for convenience and capability is *immense, complicated, and constantly changing* software on computers. This complexity and scale hinder our ability to understand the overall system, resulting in software and updates containing inadvertent mistakes, called *bugs*.[10]

The computer industry averages one to twenty-five bugs per 1,000 lines of computer code in delivered software.[11] With modern operating systems containing 45 to 100 million lines of computer code, they may have 45,000 to 250,000 unknown bugs in the operating system alone.[12] This estimation doesn't include the other programs, drivers, and plug-ins owners install in a typical computer. The result is a large number of unknown and constantly evolving bugs.

Many software bugs are insignificant or cause minor annoyances while others enable hackers to perform brutally malicious acts on computers. These latter bugs may allow malefactors to intentionally read private information, disable the computer, or install and operate their own applications without an owner's permission or knowledge. The bugs that allow another person to use the computer for their own advantage are called *vulnerabilities*. A vulnerability may be a "Weakness in an information system, system security procedures, internal controls, or implementation that could be exploited or triggered by a [threat]."[13]

Vulnerabilities may also exist as the result of poor management or ignorance by the people who manage and configure systems. A good example of this is the use of *default passwords*. When a consumer installs a new wireless router or television on a network, these devices come with default passwords preinstalled in them at the factory. Manufacturers install default passwords to simplify installation and troubleshooting problems. These same manufacturers often publish their manuals on the Internet, providing these default passwords to the public, including threat actors. Most responsible manufacturers instruct owners to change passwords when they install a new device. However, hackers have searched through installation manuals and compiled lists of these default passwords. When they encounter a device on a network they typically try the default password from their lists, and if successful, take control of the device if the owner overlooked changing the password.

Cyberspace

A third vulnerability exists when malicious individuals use valid tools and technology as designed and constructed, but for malicious purposes. An example is hackers sending a large number of emails or other messages to overwhelm a recipient's system. Another example is establishing a connection to an organization's document storage system and copying, encrypting, or erasing the files on the system. These vulnerabilities may not be eliminated within the system, but may be mitigated through other means, such as reducing their impact (e.g. maintaining file and system backups) or using more secure system controls (e.g. requiring special credentials to modify critical files or systems).[14]

An organization's legacy applications may also represent a serious threat to an organization. Many computer applications have existed for decades, and function today as they were designed. Organizations rely on these core applications to achieve their goals. Developers "wrap" these legacy applications with modern network capabilities to enjoy the benefit of providing them online. However, the original engineers who developed these applications may not have designed them to be exposed to a hostile network environment. These applications may be beset with an unknown number of vulnerabilities that are cost prohibitive to identify and remediate.

The rapid change of technology, volume of computer code, number of devices, interdependence of software, and interdependence of devices creates dynamic systems of systems, where even expert developers and security engineers are unable to predict how a computer will perform over time. This ever-changing nature of modifying and updating systems further increases complexity and disorder in systems,[15] and the resulting vulnerabilities.

These vulnerabilities collectively form the *attack surface* a threat actor can use to enter or manipulate a system. A vulnerability may act like a secret key to a door into a computer. In other words,

> Every successful cyberattack is the result of people having made the deliberate decision to connect vulnerable and exploitable computers to a network inhabited by people who have competing goals, agendas, and practices.

hackers don't create doors into computers; hackers take advantage of the ones vendors and system owners unwittingly placed there. To make matters worse, as we mentioned earlier, vendors unwittingly built these computers with the ability to allow attackers to push new software onto a computer, often without the user's knowledge. This may allow an attacker to install their own commands or software on a victim's machine, and then control how the system operates.

Some security practitioners pronounce that vendors simply need to use more secure software development practices. However, creating computer programs today is very complex, and *scientists have mathematically proven it's not possible to build a program to find all the bugs in another computer program.*[16]

While the people who develop complex computer codes are ultimately responsible for the bugs, they are prevented from eliminating many of the vulnerabilities by the complexity of the problem and drive for profitability.[17] The reality is that functionality and speed-to-market almost always trumps security. Many of the more secure software development practices, such as managed programming languages that eliminate many vulnerabilities also take a significant number of decisions out of the hands of developers and the vendors. This may limit developmental creativity, sacrifice new functionality and technology, and result in increased development costs, loss of market share, and reduced profits.

But it gets worse. Let's say developers very carefully create computer code by hand and verify the code is bug free (this would be very expensive). In 1984 Ken Thompson pointed out that no amount of code verification could ensure there wasn't malicious functionality in programs. In his landmark presentation he noted that all of the tools used to develop software could contain malicious functionality that could further embed this functionality into new systems. The only way to be sure software can be trusted is if the developers write all the code themselves, verify the code, and did the same for all tools they used to develop the code.[18]

Cyberspace

Developers and security researchers can discover some bugs in computer code using techniques such as *fuzzing*. The developer monitors an application while presenting it random or invalid data. If the application enters an error routine, crashes, or reacts unexpectedly the engineer can investigate for a bug. Fuzzing helps developers discover new instances of certain classes of bugs but will not discover or eliminate all bugs. Likewise, using teams of hackers to penetrate systems may discover vulnerabilities but will not guarantee the system is "secure" from all bugs. New bugs could result in new exploits that the developer and security industry are unaware of, and therefore have no defense against. The security industry refers to previously unknown exploits as *zero-day exploits* since the community has had no time to prepare their defenses against them.

The volume of vulnerabilities that exist today is partly the result of a rush to "network enable" homes and offices with new technology.

> Markowitz's Law states that a minimally complex system has fewer attack surfaces an attacker can exploit.

Consumer demand for devices that improve efficiency and convenience has resulted in the growing universe of the "Internet of Things" (IOT). Vendors are connecting everything from power meters, refrigerators, thermostats, wearable fitness trackers, and pacemakers to networks, often without consideration of the threats that exist, and are then targeted by malicious actors.[19] Today we see aircraft, self-driving automobiles, satellites, electrical power systems, and military weapons systems connected over public information networks.

In addition to software vulnerabilities, computer systems may also contain hardware vulnerabilities that can also be exploited. A uniquely different aspect of hardware vulnerabilities is in the difficulty that exists in mitigating them. Engineers can resolve many software bugs by rewriting or adding new code to a program. However, a weakness in hardware may continue to exist until the hardware is replaced. In late 2017 several security researchers discovered an exploitable weakness in Intel computer chips that became known as 'Meltdown'[20] and 'Spectre'.[21] The weakness could enable a hacker to steal information from the central segment of the computer system. Because these chips were critical to the functioning of the motherboards of millions of computers, replacing them was not a viable option.

The U.S. government has been one of the most vocal evangelists regarding the threat posed from digital criminals and foreign spies. And yet the vast majority of the U.S. mission-critical and national security systems operate on vulnerable commercial computers. The government's demand for low-cost commercial systems has kept demand low for secure systems, and therefore cost per unit high, further reducing demand and adoption of more secure systems by public and private users.

Space and Time

The geometry that exists in the physical domains of war is well defined. We can measure how far and how fast a plane flies, the depth a submarine dives, or the distance between satellites in space. This geometry directly impacts the concept of time when applying force.

The application of power in traditional warfare is bounded by a nation's ability to project force. This ability to project force on land, sea, air, and space is directly affected by distance and indirectly by time. For example, the time needed for the United States to project force on the Korean Peninsula is small due to the relatively short distance prepositioned forces need to move. Contrast that to the amount of time it took the United States to project force to engage in combat operations in Afghanistan in 2001. In this latter case, it took months to transport fully equipped forces into that land-locked country.

A secondary corollary of the geometries of land, sea, air and space is how they are bounded to national identities. Countries such as Iran or Venezuela may desire to impose their political will on other nations. However, overtly violating their potential victim's national boundaries may result in unacceptable costs.

Cyberspace operations are not bounded by geometry, and states view cyberspace as an economical and reasonable place to project power and impose their political will on their adversaries. It is unreasonable in today's world to expect that a person could travel hundreds or thousands of miles, breach a security perimeter, execute their malicious goals, and extract themselves from the target area in seconds. However, this is precisely what occurs hundreds or thousands of times every day around the world via cyberspace. The speed of communications across the Internet enables a nation-state, cyber-criminal, hacktivist, and other malicious actor to extend their reach hundreds or thousands of miles in seconds.

The sensitive government documents released by Edward Snowden alleged the United States has a willingness to breach the geographic geometries of other nations in cyberspace. These disclosures highlight just one nation's efforts to exploit the speed of action that can occur in cyberspace. McAfee noted that as early as 2007 there were 120 countries, including the United States, using cyberspace to target financial markets, government systems, and national infrastructures.[22] It is logical that today every country would leverage cyberspace to achieve their political, military and economic goals.

This isn't to say that geography doesn't matter. Cyber evangelists advocating for government sponsored offensive cyber operations contend that it is difficult for a cyber-attacker to determine the geographic location of their target in cyberspace, and therefore identify the national sovereignty they may violate in a cyber operation. Regardless, all computing devices, and many communication mediums, are still physically within some nation's borders, and therefore susceptible to that nation's laws. A cyber attacker's activities may also pass cyber weapons, operational commands, flooding attacks, or spoils of war through unwitting third parties. Nations conducting these operations and negatively impacting these unwitting third parties often fall back on the "collateral damage" defense, arguing that the unfortunate resulting damages from deliberate attacks were unintentional consequences, incidental to operations.[23] These pronunciations run the risk that adversaries could use the same justification, leading to an escalation of unconstrained, destructive cyber operations.

The time necessary to make changes in cyberspace also differentiates it from other domains. Development and deployment of traditional military technology takes time, and modern battlefields are "come-as-you-are" affairs, with military forces arriving and fighting with the existing technology in their inventory. Forces may arrive ready to fight the last war and not prepared for technological advancements, such as what occurred to Iraq in the 1991 Arab-Gulf War.

Cyberspace however is marked by rapid advancements that create fundamental changes in technologies and vulnerabilities. Technological changes tend to be driven by the commercial demand from users and organizations for the newest capabilities. Vulnerabilities and exploits evolve very fast, with tens of thousands of new and modified software, malware, exploits, and patches released every week. This constant change leads to a highly dynamic environment.

Security Controls

The Internet began as a research project to enable one computer to send any data over commercial telecommunications media to another computer. Since the researchers trusted each other security was essentially an afterthought. Additionally, since the Internet today is a collection of networks and organizations, there isn't any centralized management or control. The result is a resilient and reliable information dissemination system. However, the Internet also does not provide any endemic security. Network and system owners must add features, often referred to as *security controls*, to protect systems and information.

Individuals and organizations deploy security controls to prevent, detect, and respond to unauthorized and malicious behavior. We noted that scientists have proven mathematically that it is not possible to produce a general-purpose security control that can detect all malicious functionality in computer code. In response, the majority of security tools, including anti-virus and intrusion detection suites, are based on *retrospective analysis*. Many of these tools can only detect attacks that researchers have previously identified, analyzed, and added to their databases of capabilities. New threat techniques or vulnerabilities, or the slight modifications to existing threat techniques or vulnerabilities, will slip past these security tools until each specific instance is added to the tool's database.

New technologies, such as "machine learning" or "artificial intelligence" have been advertised for decades as a means to detect innovative or new attacks. However, most of these technologies have struggled to correctly learn to identify anomalous activity from normal behavior, or bad activity from good activity. These techniques, when created in a lab environment, are very capable of identifying new attacks if they differ from the static environment. However, they struggle when taken from the laboratory and placed into operation. Two direct factors that affect these technologies are the presence of a realistic and clean training environment, and dynamic operational environment. These tools must be placed on a clean environment to learn "normal." However, if the adversary is already in the environment, the tools will record their malicious activity as "normal." Second, if the environment or user behavior changes due to organizational changes or even severe weather the intelligent tools will flag these changes as anomalous. These tools also end up requiring retrospective analysis to update their configurations. An additional challenge to learning "normal" is that many computing actions occur very infrequently, for example, once a month or once a quarter. Machine learning is challenged to "learn" valid from invalid actions when the periodicity of actions changes.

Since most security controls require retrospective analysis, over time these controls grow more complicated as vendors update them to deal with newly discovered vulnerabilities and exploits. A recent government study showed that while the average defensive tool complexity has grown nearly exponentially, the average attack complexity has remained constant. Security tools have grown so complex that nearly one-third of all Department of Defense mandatory security updates involved the security tools themselves, meaning that the tools meant to protect networks introduced vulnerabilities into the network.[24]

As an example, in May 2014 Symantec, the makers of Norton Anti-virus, noted that sophisticated attacks will compromise a computer protected by anti-virus tools, and that these detect at most 45% of current malware.[25]

Security vendors have responded to increased demand for security by developing new and custom technology that addresses individual threats. The security community uses the phrase "defense-in-depth" to refer to the elaborate set of tools available to protect a network or device and detect and respond to compromises when prevention fails.

Those familiar with the military idea of defense-in-depth should not confuse the kinetic and the cyber security versions of the concept. In kinetic warfare defense in depth refers to layering defenses linearly to delay and attrite attacking forces, reducing an invader's effectiveness as they penetrate deeper into defenses. In the cyber realm it refers to layering defense technology in the hope that one of these technologies will discover malicious information or code and prevent it from causing damage. These defenses don't attrite an attacker, since the attackers are free to continue attacking this or another target.

> *Defense-in-Depth* means layering a variety of security controls in series, e.g., firewalls, intrusion detection systems, and antivirus.
>
> *Defense in Breadth* means layering a variety of similar controls in parallel to improve effectiveness, e.g., multiple versions of antivirus products on a system.

Today the concept of defense-in-depth has taken on a life of its own. The NIST Risk Management Framework lists hundreds of security controls that the U.S. Government recommends defenders install. The framework does not provide a justification for individual tools, or establish metrics that demonstrate the effectiveness these controls will have on attackers, or how implementing these controls may impact operations or other controls. The result is that the security community is putting tremendous effort and funding into making security controls bigger and more complex.[26]

A 2013 study found that layering security controls did not guarantee improved defense effectiveness. The researchers compared 37 security products from 24 different vendors to detect and defeat 1,711 real-world exploits. The researchers found that "none of the 37 security products detected all exploits, and only 3% of 606 unique combinations of two security products detected all exploits. Further, [they discovered that] there is a large diversity in the security performance between individual security products or combinations of security products."[27]

Increasing the quantity of security controls may create additional pitfalls. Adding controls brings the need to maintain and operate the controls. These additive controls may overwhelm a security staff unless management also increases the resources necessary to maintain and operate the controls. This may create a situation where more controls result in a less secure environment and creates a false sense of security. This is especially true if the staff consumes their time and resources maintaining controls and not maintaining their systems, thus not mitigating known vulnerabilities. Additionally, an organization could turn the process into a "checklist" mentality where the required controls are installed, but not properly configured, managed, or maintained.

The increase in the number and complexity of security controls may result in a more complex overall system and a commensurate increase in the organizational attack surface. Security controls are themselves computing systems that often operate at a higher level of privilege and subsequent trust level. As noted previously, their increased complexity and therefore potential for vulnerabilities, combined with operating at a higher trust level, results in the potential for an attacker to compromise a system by attacking vulnerabilities in the very components deployed to protect a system.[28]

Dependencies

Defenders are often surprised to discover relationships between their kinetic operations and their cyber services and technologies. During 2001 the Code Red worm ripped across the Internet infecting the Department of Defense's (DoD) Microsoft servers and flooding the DoD Internet access points with scans looking for machines to infect. To prevent the scans from flooding their internal network and degrading services the DoD blocked all incoming web traffic. Numerous DoD operations were affected, including shutting down traffic on the Mississippi River, since the Army Corps of Engineers used an Internet-connected system to control the locks on the Mississippi River.[29]

Attacks that degrade or damage systems may have similar unintended consequences. If it's difficult for defenders to understand the dependencies between their own technologies and operations, imagine the difficulty for attackers to understand the consequences of using offensive cyber weapons.

A cyber-attack could digitally decapitate a nation if it removes the ability of the victim's leadership to communicate and maintain control of their military forces or their nation. This decapitation may prevent a leader from suspending military actions resulting in protracted violence. Segmenting or disabling communications may prevent leaders from communicating with forces to tell them to suspend hostilities or disable weapon systems and may be mistakenly seen by observers, opponents, or the populace as a sign of further aggression or unwillingness to capitulate resulting in a protracted and possible escalation of war.

Psychological Effects

Some discount cyber-conflict since it is not warfare in the traditional sense of the word. These traditionalists may define warfare as a state's use of physical violence to achieve a political goal. However, a dependency on the digital domain, and the potential effects of a cyber-conflict may be enough to force a decision-maker to capitulate. Richard Clarke, the former National Coordinator for Security, Infrastructure Protection, and Counter-Terrorism noted that our national reliance on cyberspace may be enough to enable a foreign threat to deter the U.S. from taking some action, or compel the U.S. to stop performing some action that is in their best interest.[30]

Kinetic warfare uses physical effects or the threat of physical effects to generate psychological impacts. In some instances, a cyber-attack that damages or degrades personal or financial information may also create psychological effects without the kinetic action. Alternatively, cyber-attacks that impact physical control systems used to maintain power, water, or sewage infrastructures may create physical effects that consequently generate psychological effects.

It should be noted that financial or political "elites" in the "First World" are much more reliant on cyberspace. These elites may realize larger consequences to a cyber-attack, since damage to stock markets are not much of a concern to someone who doesn't own stocks, even though there may be cascading effects that eventually impact them. Similarly, a poorer country may view the launching of a cyber-attack as a means to impact richer nations that rely on cyberspace while they are not as significantly reliant.

The consequences of a cyber-attack may not be as significant as an attacker initially believes. Many humans are resilient and will adjust when faced with adversity. Likewise, when faced with degraded or loss of computing and telecommunications systems and the benefits they provide, victims will look for alternative ways to achieve their goals. A massive power outage as the result of a cyber-attack would have major impacts to commerce, health and safety. However, people and organizations would adapt and continue living until power is restored.

The psychological effects of cyberspace should not be discounted. The anonymity of cyberspace enables many residents to lash out. For some cyberspace provides a podium for anonymous attack against those with whom they disagree. For others it provides a vector to spew vitriol they may never condone in person. Vicious personal attacks once placed online, and the deep personal damage that results, may be difficult to erase. Once the realm of bullies and cyber trolls, nations have also jumped in to manipulate personal opinion, as with the apparent Russian manipulation of the US election.

Decision Making

People make better decisions when they have the appropriate facts and can conduct informed analysis on the costs and consequences of their actions. However, cyber defenders often do not have the information necessary to make informed decisions. The defenders may have little to no understanding of the real (and undiscovered) vulnerabilities in their system, their adversaries' capabilities, the effectiveness of their security controls, and the consequences of their own actions. While there is a very robust (and vocal) risk management community, the processes they profess appear closer to wishful thinking wrapped in engineering terms and structure.[31]

The attacker on the other hand, can gather information on their target's people, technologies, and processes. They can develop broadly scoped and explicitly detailed profiles of their intended victims, discover the systems these targets use, research relevant vulnerabilities, and rehearse attack tools and techniques on their own systems until they are ready to launch their attacks. All of this provides attackers with significant advantages.

Additionally, when either an attacker or defender makes decisions in cyberspace, their actions may provide actionable information to adversaries and their systems. For example, if a defender detects an attacker in their systems they may decide to disconnect affected systems to remove the attacker's access from these systems. This action may unintentionally announce to the attacker that the defender detected them. The attacker could embed code in their attack tools that detects when the compromised system can no longer communicate with the hacker, erases itself and perhaps destroy the victim's information and systems to hide any evidence of their nefarious activities. In each of these instances, the attacker has relevant information to modify their attack methodology to maintain their freedom-of-action.

Cognitive and social scientists understand that people have *cognitive biases* that lead to incorrect and ineffective decision making. There are numerous causes for this poor decision making including using rules of thumb, analogies, emotions, and social influences.[32]

Cyberspace

Table 1 - Example Cognitive Biases

Cognitive Bias	Description	Cyberspace Examples
Dunning Kruger Bias	Causes one to be unaware of their performance – and overestimate their competence.	Empirical evidence demonstrates many cybersecurity professionals overestimate their ability to protect an organization's information and systems, especially against motivated, resourced, and skilled adversaries.
Functional Fixation	Limiting the use of a tool to the way it was designed.	In 2011 one-third of the U.S. military's most critical vulnerabilities were in security controls – hackers could attack these security controls, which may not be maintained or monitored for attacks, to obtain access to other systems.
Confirmation Bias	Looking for information that supports an existing hypothesis while ignoring evidence that invalidates a hypothesis.	Security testing is usually conducted in an attempt to demonstrate systems are secure, not insecure.[33] Users may perceive system security is greater over time if there are no observed malicious or negative effects on the system, although this may only be an indication malicious activity occurred but was not detected.
Tragedy of the Commons	Selecting locally optimal choices that create globally worse results.	It may not be rational to consume resources to filter traffic leaving a network since this doesn't improve that networks, but may reduce malicious actions to others.
Cognitive Dissonance	Tension results from two conflicting thoughts or behaviors. ex: "everyone is out to get you" or "they are evil."	Some cyber defenders argue that conducting cyber intelligence operations is putting everyone at risk, when in reality intelligence operations are performing a very important national function. Some argue, "[all] hackers that expose vulnerabilities are evil."
Ingroup Bias	Viewing "their" group as better, while outsiders are collectively viewed as inferior.	Perception that foreign intelligence collection is evil, while a person's own nation may be conducting the same operations against others.
Cultural Bias	Interpreting and judging a situation by standards inherent to one's own culture. (Wikipedia.org)	Western military thinking tends toward Clausewitz (defeating military forces) while eastern military thinking tends toward Sun Tzu (achieving objectives without warfare). *
Survivorship Bias	Concentrating on the people or things that "survived" some process and inadvertently overlooking those that did not because of their lack of visibility.	Many people look to governments for leadership in cyber-conflict, when the news media regularly report that government systems are not maintained and are often compromised by unsophisticated hackers.

* Clausewitz and Sun Tzu are important historic military strategists; however cyber strategists must be careful to avoid blinders that force their observations and theories to match century old thought.

A simple example is when people force cyberspace operations to match kinetic operations. Kinetic military terms such as *warfare*, *defense in depth*, *kill chain*, *high ground*, and *active defense* may have similarities to cyber operations. However, these terms may not have a similar meaning in cyberspace operations, and trying to force the military aspects to fit cyber may result in ineffective decision making at best and wishful thinking at worst.

The ambiguity and deception enabled by cyberspace may increase the potential for ineffective decision making for both the attacker and defender. However, attackers have a shorter decision-cycle and may acquire and exploit information with greater agility, possibly giving them an advantage to conduct their missions and achieve their goals.

Privacy and Anonymity

Privacy and anonymity are different but related concepts. Privacy involves keeping the content of information from being exposed to others. Anonymity involves keeping the identity of someone responsible for an action or message from being exposed. The approach to privacy and anonymity as human rights differs considerably across geo-political, national, and cultural lines.

Privacy and anonymity may be viewed as fundamental human rights in some nations, while others put the needs of the nation ahead of either. Likewise, some states also fervently protect their nation's and citizens' privacy though laws, security clearance processes and national security acts.

All things considered, there are still a great number of actions in cyberspace that necessitate the ability to attribute an action to an identity. Despite the need to ensure privacy in cyberspace, information systems often need to attribute a unique identity to their authorized privileges before authorizing the identity to perform some action. Online shopping is a prime example of this condition. Additionally, after an event occurs, an analyst or investigator may desire to attribute malicious actions to the individual(s) responsible.

Many networks intentionally but inadvertently complicate attribution of identity[34] by using technology such as port or network address translation (PAT or NAT). These management tools shield an internal network addresses from an external entity.

For example, an Internet service provider (ISP) may assign one public IP address to a home network, and the home network assigns private IP addresses to the computing devices on the network. Messages appear to be coming from the single IP address associated with that house network. Since all the internal traffic leaving the network has the same external IP address, it's difficult for someone outside of the network to determine which machine originated an attack — mom's iPad, dad's desktop, the kid's game console, or a neighbor that illicitly connected into the home wireless network.

Dynamic Host Configuration Protocol (DHCP) is another tool that network administrators use to manage computers' addresses by assigning different IP addresses every time a device connects to a network. Most service providers, broadband home networks, cyber cafes, and large public and private networks use DHCP to automate the management of networks. Since these technologies hide and change IP addresses, they unintentionally make tracking and identifying activities in cyberspace difficult and unreliable.

The use of DHCP as an inadvertent 'weapon' of anonymity is one that hackers routinely take advantage of, purposefully or not. Researchers from the Ponemon Institute released their 2018 'Cost of a Data Breach Report' where they show that the mean-time to detect a compromise was 196 days, and the mean-time to contain a compromise is an additional 69 days.[35] By the time a hacker's activities have been discovered in a defender's network the hackers have disconnected from the network, and the IP address the hacker used has been returned to the pool of available addresses. Additionally, this IP address may have been reassigned to other systems possibly several times. Most home and corporate networks are not configured to maintain logs of IP address assignments, nor log sufficient information to identify a hacker over a reasonable period of time.

Worse yet, it is trivially easy to obtain unattributable access to cyberspace. Coffee shops, airports, libraries, hotels and schools often

> The fundamental paradox of cyberspace: The technology that makes our life easier, also makes an attacker's life easier.

offer free wireless Internet access points. Municipalities and private enterprises are also making efforts to offer free, anonymous wireless and WiMAX[36] access.

Some organizations make efforts to hide the content of their information. Encrypting communications may delay an adversary from learning the content of communications, but may inadvertently identify that the encrypted information, and the people with whom the communications are taking place, are of value. Indeed, just discovering *metadata* [37] that a group of people are sending messages between themselves reveals that the individuals have a relationship.[38]

Additionally, Internet services may permit people to establish email addresses, websites, social networking accounts, chat accounts, peer-to-peer accounts, and other access without validating their actual identity. Other services allow individuals to traverse the Internet and many of its functions anonymously. A prime example is the *Tor* network. Tor is a self-contained network that operates in a peer-to-peer manner, shielding the entry/exit points and the origin and destination addresses of the messages traversing it. Malicious actors can leverage Tor to hide their activities from detection and backtracking.

The result is that attackers can make their attacks appear to come from nearly anywhere or anyone. They create fictitious e-mail, social network, and financial service accounts, and false personas to hide their true identities. Likewise, they hide their actions using anonymous networks, proxies, and hop points in multiple countries to mask the originating computer and address. The combination leveraged against varying and complex national laws allow attackers to remain anonymous.

Today online fraud is a growing business, affecting everything from online purchases to banking. Fighting fraud is a balance between providing utility to customers and preventing losses. Businesses are beginning to have success combatting fraud by analysis and verification of customer-provided information including address verification services, credit card verification, device fingerprinting, fraud scoring models, and enhanced payer authentication.[39]

Cyberspace

Just one example of fraudulent activities are the scammers prowling personal dating sites using personas and photos harvested off social media. These criminals have been particularly successful posing as single military service members assigned overseas to exploit naïve 'marks', breaking hearts and defrauding them of their money. This occurs due to the ease in creating false personas and an inability to confirm or deny a person's online identity.

Deception is not limited to the attacker. Defenders can create false networks, called honeypots or honeynets, to bait attackers and learn about their tools and techniques. Similarly, defenders can configure computers to respond differently to apparent hacker probes to confuse and delay attackers. Defenders can also embed *beacons* into their data that call home when the files have been illicitly taken and accessed, announcing information about the perpetrators, including their location.

A successful compromise of a public or private organization's system may create major risks to the organization's livelihood. The admission of a breach may result in internal and external investigations, civil or criminal actions against individuals, and loss of future business.[40] Government victims face the possibility of a government audit or an Inspector General investigation into the root cause and responsibility of the breach. Commercial organizations may face intense scrutiny by regulators, the public, and shareholders. Corporate sales dropped 2.5% in the fourth quarter of 2013 following the December 2013 credit card breach at Target, in part as a result of the breach. These potential impacts may lead some organizations to hide breaches to prevent the scrutiny, embarrassment, or loss of jobs. Several former executives of Target Corporation and HB Gary may attest that public knowledge of the breaches against their companies contributed to their release from the companies.

Victims that choose to hide incidents potentially deprive others knowledge about the attack, methods to detect and mitigate the attack, and steps to protect others from also falling victim to the threat. On March 1, 2018 Equifax reported a compromise of 143 million American's sensitive personal information. Allegedly, the breach was discovered in July 2017.[41] In the case of the theft of customer data, delaying victim notification gives the perpetrator more time to exploit the stolen data.

Trust and Distrust

As Metcalfe notes, computers connected via telecommunications media create greater utility. However, collaboration requires trust. Organizations often establish *transitive trust*, whereby organizations that trust each other establish procedures to access systems and share information for mutual benefit. However, enabling electronic trust may allow a malicious individual to use vulnerabilities on a trusted partner's network to access the other system.

Also, trust can be breached by a vendor, software manufacturer, or even an insider who establishes a *back door* into a network. A *back door* is a mechanism that someone creates to gain access to a system or software without using standard procedures or protocols. Software manufacturers have done this so they can fix or update their software without their customer knowing that a problem exists. Individuals create back doors into a system or network segment for much the same reason as vendors. These back doors are typically established in a way that bypasses established security protocols.

Attackers can also exploit the inherent trust network administrators and users have in their data. Attackers can embed false information in tools, commands, and other evidence to confuse analysts. Attackers can browse sites with their browsers set to foreign languages or by embedding foreign commands into attack tools to cause analysts to believe the visitors are from other countries. One of the most famous Internet attacks was the 2010 Stuxnet attack. In this case malware allegedly infected Iranian uranium enrichment centrifuges. Journalists reported that Stuxnet was a joint U.S.-Israeli campaign, although there was no evidence in the Stuxnet code that pointed to this conclusion. The evidence was initially based on anonymous reports, opinions, and speculation.

Legal and Cultural Complexity

Cyber-attackers and defenders have learned to exploit the presence and absence of legal agreements between nations. In the United States legal entities like the United States Secret Service and the Federal Bureau of Investigation are more likely to receive requests for assistance from countries such as Canada or the United Kingdom than from Iran or Venezuela. Countries that are hostile to the United States could openly deny assistance to another government's investigation into an attack. Threat actors can leverage this knowledge by using hostile nations as 'hop points' to help establish and maintain their anonymity, hopping through several countries known to be hostile with the target country. Equally frustrating to defenders are indifferent nations that either lack the resources or choose not to apply resources to aid in an investigation of a cyber-attack. We often see this indifference in third world and developing countries.

Attackers use compromised systems around the world as 'hop points' in their attacks. These computing platforms fall under the legal framework of those countries or principalities. Many of these countries have neither effective laws that adequately identify certain activity as a crime, nor the requisite skills, capacities or resources to investigate compromised systems that may only be a hop point of an attack. The more sophisticated the attacker, the more likely they are to use multiple hop points when conducting their activity. This enables attacks to achieve a high level of anonymity.

Attackers can also leverage companies around the world that provide Internet services, including Internet access, web hosting, email hosting, cloud hosting, and social networking. Some of these companies are nationalized, some quasi-governmental, others purely commercial. Each carry with them unique philosophies regarding participation and anonymity within the Internet. Additionally, they may share similar cultural beliefs and attitudes with the central government of the country in which they reside. Because of these cultural beliefs, they may aggressively combat efforts to determine who their customers really are, thus providing attackers with an active ally against investigators.

Alternatively, federal governments may require commercial service providers within their borders provide their customer's identity, locations, data, communications, or behavior. United States federal laws require companies operating within the United States to provide information to law enforcement when demanded through legal processes such as warrants and subpoenas, including National Security Letters.

Vulnerability Availability

An operating system (OS) is a core application and supporting applications that manage the hardware configurations, software interactions and communications with a computer. Depending on the platform, the OS may contain a few million to tens of millions of lines of code. This volume of code provides malware writers a tremendous number of opportunities to identify and exploit vulnerabilities in the system to gain full control of targeted systems and the information it contains.

Unpublished *zero-days*, are highly prized in the security and hacker worlds. The most prized vulnerability is one that allows an attacker to gain 'root' or system access when properly exploited. Most zero-day vulnerabilities that are exploited provide attackers with a head start in gaining access to, and further exploiting a targeted environment.

Recently, developers have taken steps to engineer their operating systems more securely, as well as more rapidly patch known weaknesses in their software, reducing the window of vulnerability exposure. This has induced attackers to target the other applications that run on computing platforms. Attackers have increased their focus on third-party applications that are most prevalent across all operating systems, such as Java, Adobe Flash, and Adobe Reader. These applications are typically found across Apple, Linux, and Microsoft systems. The successful exploitation of a third-party application allows an attacker to gain their initial foothold in the target environment. With this foothold, they can broaden their access to that system, as well as move across the environment to other systems.

In addition to broadly deployed third-party applications, attackers may also focus on a specific company or industry, deploying an exploit that is crafted for a narrowly deployed application. These unique software targets lack the breadth of deployment, and as a result, may be more challenging for security experts to identify. A clear example was the targeting of Windows-based applications developed by Siemens Corporation for industrial control system (ICS) functions. The Stuxnet worm was specifically crafted to exploit the Siemens ICS applications, thereby affecting thousands of ICS installation deployments.

Many specially crafted applications, such as customized web applications, may only have one or a few versions of the software deployed across the Internet. Organizations often contract the development and update of these types of software to companies that specialize in this work. These applications may present the opportunity to exploit a weakness with the advantage that broad security analysis may not discover and announce the weakness. Due to the difficulties discovering vulnerabilities in these custom systems, successfully targeting these applications may require a more patient, resourced and sophisticated threat.

Successful attacks do not necessarily need a zero-day exploit or an undetectable piece of malware. Many organizations build and modify their internal networks to enhance the functionality of their supported audience (e.g. employees, students, vendors, guests). Attackers may leverage existing file-sharing services to gain access into the network. Once inside, they can capture administrator rights to install software packages to broaden their foothold inside the target environment. They then traverse the entire network searching for available information to capture and exfiltrate. Indeed, attackers leverage the fact that most security teams focus their attention on network traffic crossing an Internet boundary, rather than originating from one internal network segment to another, or from one internal host to another. Attackers may use common scanners to 'map' an organizations internal network for later exploitation. This inherently built-in functionality in modern networks is often the most efficient means for an attacker to conduct their operations while reducing the possibility of discovery.

Sophisticated attackers will also search for the domain administrator's credentials to enable them to remain on a victim's systems, remaining virtually invisible to defenders. Hackers also attack personal devices, children's devices, online games, and networked building systems such as heating, air-conditioning, telephones, and elevators. That is, they use tradecraft rather than zero-days.[42]

Weapon Availability

In traditional warfare there are resource limitations that restrict the type and volume of weapons and munitions military forces can bring to bear against an adversary. A clear differentiation between cyber and kinetic conflict is that the available cyber weaponry is essentially infinite. An attacker may develop an effective piece of code and reuse it an infinite number of times. A minor or major upgrade to this weapon is only constrained by the attacker's imagination and time to invest in the modification. When completed, an attacker can deploy a nearly infinite number of this newly evolved weapon. This contrasts significantly from the defender who leverages a finite number of increasingly complex defensive tools to combat the continuous barrage of attacks.

One need only enter the phrase "*Open Source Hacking Tools*" in an Internet search engine[43] to discover millions of pages of information and repositories for uncompiled[44] tools that range in functionality and sophistication. These repositories operate in the same way as other open source projects. Development and enhancement of some tools are crowd-sourced by like-minded people, creating software packages that meet the developer's goals. With the source code uncompiled, it remains difficult for anti-virus companies to search for and develop signatures against this code, since the code can be altered to bypass security controls when compiled. The amount of code that can be used in an attack, which is developed on a daily basis, is staggering. Some estimates are in the millions of new pieces of malware created daily. This smorgasbord of open source and freely available malware adds to the complexities that face defenders and aid attackers.

In addition to open source applications, individuals, groups, and companies develop software both commercially and on the black market. One of the most prolific commercial hacking tools has been BackTrack.[45] Many information security professionals refer to BackTrack as the Swiss Army Knife of hacker tools due to its varied functionality. On the other end of the spectrum, malware like Zeus was created and sold in black market forums. Zeus was an information stealing toolkit that allowed attackers to capture login credentials for email, social networking, banking or other functions. Zeus was so successful that the developer became a prime target for law enforcement officials, eventually resulting in the arrest of Hamza Bendelladj in 2013, and his subsequent extradition and conviction in the United States.[46]

An even darker threat to organizations with highly valuable intellectual property or intelligence is the availability of covert groups and individuals that develop customized cyber weapons for a price. Similar to the commercial tools sold in black market forums, highly skilled and intelligent code-writers will develop customized exploits, Trojans, or various other malicious code. This code will target applications known to exist in the target's environment and will not be susceptible to detection by security controls. The malware may bypass all existing security mechanisms in the target organization.

Some attackers outsource components of an attack. One actor may acquire ten user logins, the next actor will install a remote access Trojan on five of the systems, and another establishes a remote access Trojan on the other five systems. The segregation of duties allows the owner of the attack to manage the various workflows without putting the operation at a high risk of being hijacked by their "contractors."

Defenders also make the attacker's job easier when they announce the security tools they employ, often on social media. This intelligence enables attackers to purchase commercially available security controls, and then build, test, rehearse, and refine their attacks against the very security controls they will encounter on these environments.

Vulnerability and Weapon Lifespan

The exclusivity of a vulnerability is a fleeting property. A vulnerability frequently exists in systems without individuals or organizations being aware of the vulnerability. Others may be aware of the vulnerability, even if the knowledge was not made public.

The individual who discovers a vulnerability may choose to disclose the vulnerability to the vendor or the security community to have the vulnerability eliminated. Alternatively, the discoverer could decide to not disclose the vulnerability, and instead *weaponize* the vulnerability as a new, zero-day attack tool. These zero-day attack tools provide threats with the means to impact vulnerable systems with a capability that retrospective analysis is not capable of detecting or protecting against.

The risk associated with the discovery of a zero-day vulnerability, use of an exploit, disclosure of the vulnerability, and the elimination or mitigation of the vulnerability is referred to as the *window of vulnerability*.[47] The size and shape of a window of vulnerability determines the risk an organization inherits.

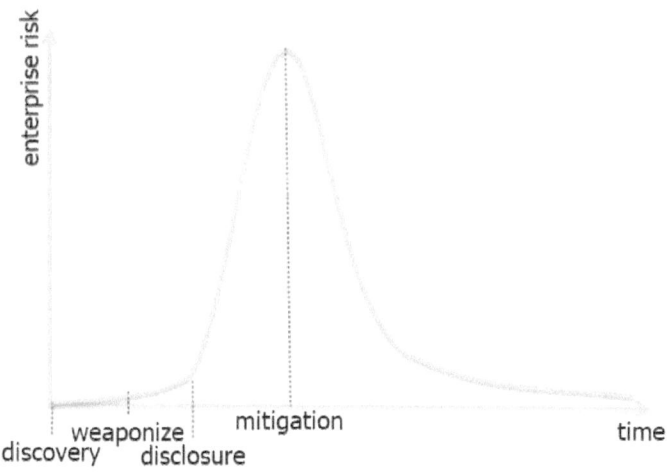

Figure 1 – An *Attacker's* Window of Vulnerability

Traditional theory is that every organization's risk increases considerably once an attacker weaponizes, and as a result, discloses the vulnerability to the public. Risk doesn't decrease until the vendor release a patch, update, or other security control that mitigates the vulnerability.

Reality is much more complex and dangerous. An attacker maximizes the victim's risk immediately after the attacker stealthily compromises their network, since the attacker may possess unfettered access to the victim's systems and information. This risk remains high until the victim discovers the attacker, removes the attacker's access, and removes the attacker's ability to reenter the vulnerable network. The risk doesn't drop significantly until the community has the means the mitigate the vulnerability.

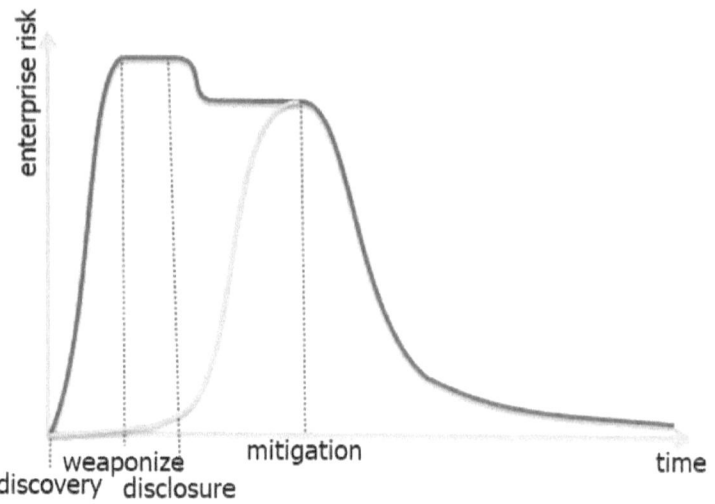

Figure 2 – A Victim's Window of Vulnerability[48]

However, when an attacker "uses" a weapon, they often transmit a copy of the weapon to a target, where that target or any intermediary systems can record the weapon. Since a weapon is composed of computer code the victim is able to record and analyze it, to discover how the weapon operates, and use this knowledge against the perceived attacker, their allies, or innocent Internet bystanders.

A zero-day weapon requires a previously *unpublished* vulnerability. This is different from a previously *undiscovered* vulnerability. While a researcher may conserve a newly discovered vulnerability for a future offensive attack, another researcher could simultaneously discover the same vulnerability and embed defensive details in security controls to detect and/or prevent the apparent zero-day from operating effectively. Use of this tactic could inject significant doubt and risk into an attacker's decision-making process.

The Human Vulnerability

Attackers recognize the frailties of users of computing systems. Similar to the Chinese military strategist Sun Tzu's principle of attacking a target's weakness rather than their strength, a common saying in social manipulation is, "You don't attack a powerful person, you attack a powerful person's person." For example, hackers may be more successful attacking a powerful person's support staff or family members, and use the established trust between the two parties to access the powerful person's information or systems.

Attackers may employ multiple means to exploit or 'trick' their targets, including:

- Phishing. Phishing is the distribution of e-mail that contains a malicious attachment or link to a webpage. These messages often exploit vulnerabilities on a computer system or seek to capture the target's credentials. An attacker exploits the asymmetric nature of cyberspace and resulting low cost to send numerous copies of messages hoping a number of "marks" will fall for the trick. Once the attacker is operating on a compromised system the attacker broadens their access and attains their goals.
- Whaling. Whaling attacks are similar to phishing attack, but target a specific, high-payoff victim with a customized message.
- Watering Hole Attack. In a watering hole attack a perpetrator compromises the website of a valid organization their targets are known to frequent. For example, if an attacker wanted to target a political party, they may compromise the website of a think tank that shares the general philosophy of the party. The attacker plants an exploit on the website that targets the computers of visitors to that site. In this attack the hacker 'lays

in wait' for their 'prey' and lets them walk into the trap, much like a 'crocodile at a watering hole'.
- Social Engineering. In a social engineering attack, the attacker manipulates marks to take actions that they believe are for a proper purpose but are unwittingly benefitting an attacker. For example, an attacker may pose as a network administrator and ask a database owner to provide their database username and password to "update the database application with a new security patch." The attacker may refer to actual people in the organization to give credibility to their request. Once the attacker has the logon and password credentials, they may now have unfettered access to the system and the data that resides there.
- Swatting. Law enforcement agencies are not immune to manipulation. Cyber-attackers have used cyberspace to research the location of potential victims, then called emergency dispatchers (sometimes with Internet-based phones) and have emergency services (e.g. Special Weapon and Tactic [SWAT] teams) dispatched to their victim's location.[49]

Though there are finite types of social engineering attacks, the individual aspect of different attackers coupled with the uniqueness that accompanies organizational environments, lends to an expansive range of attacks that have the potential for success. This ability to successfully target humans is a clear indicator of the weakness that they present to an internetworked environment.

Resilience

At the macro level, people tend to be psychologically resilient. However, warfare and many criminal acts exploit the reality that structures and people are brittle when exposed to physical violence.

In cyberspace the bugs in code, exploitable vulnerabilities, device failures, power surges and outages, and acts of nature all conspire to cause losses, outages, and system failures. However, if we think of computers as digital brains, then we see that they too tend to be resilient. Often operators can quickly repair or reload corrupted systems to restore them to a pristine (though still vulnerable) condition. Likewise, like a brain, large digital networks adapt by reconfiguring and rerouting messages around outages.

While the digital nature of information systems and networks allows them to be quickly restored, just as in the kinetic world there are exceptions and limits to this resiliency. Physically damaging devices, especially highly specialized devices like those in electrical power generation and transportation, can increase the duration and impact of incidents.

Likewise, just as in the kinetic world, the exposure or corruption of critical information can have devastating impacts to persons and organizations. The theft of sensitive information, such as personal information, corporate proprietary information and trade secrets, and national security information may have lasting effects.

This chapter provided the current nature of cyberspace. An understanding of this *engineered environment* is necessary to understanding the *principles* of operating within this environment that we propose in the next chapter.

Cyberspace

Chapter 3. Principles of Conflict in Cyberspace

The U.S. Department of Defense declared cyberspace a new warfighting *domain* that complements the existing land, sea, air, and space domains.[1,2] It's natural for warfighters to believe the principles in these domains also apply to cyberspace. However, unlike the other domains, cyberspace is a man-made environment that differs significantly from kinetic domains.

Additionally, there are no uniformly accepted principles of warfare. Warfare is state-sponsored violence or threat of violence in the pursuit of a state's political goals. Nations both implement and constrain this violence based on their societal norms, the nation's experiences in warfare, and their particular military culture. It should be no surprise that principles of war differ across nations based on these national distinctions.

Since many nations have transitioned aspects of political, economic, and military power to cyberspace, cyber-conflict is now affecting all aspects of national power. A U.S.-centric view of cyberspace might argue freedom of expression, anonymity, and privacy are key concepts. More repressive states may have a more restricted view of cyberspace. The result is a situation where citizens and governments from nations with differing political and societal ideologies view conflict in cyberspace from differing perspectives.

> These conflicting perspectives may create *security dilemmas* where a state's actions to increase its security, such as collecting intelligence on potential threats or economic opportunities leads other states to react. This increases tensions and escalates conflict. This may occur when neither side desires conflict and the conflict may harm all parties involved or affected.

Principles of Cyber Conflict

However, states are not the only actors contributing to cyber-conflict. Crime is any action violating the laws of a state. In many cases it's not the effect that causes a state to declare an action criminal or an act of war, but the perpetrator's intent and whether a government authorized the act. For example, governments allow their proxies (intelligence organizations, police, military, etc.) to commit violence or suppress rights (conduct surveillance, execute searches, etc.). These acts cause significant challenges in cyberspace where it is difficult if not impossible to identify if a state or some other entity was responsible for an act, let alone determine the intent of the act.

Despite these challenges, there are fundamental principles to conflict in cyberspace regardless of the entities performing the actions or their intentions.

Anonymity

Anonymity is a key principle to cyber-conflict and is a central or supporting component to several other principles. Sophisticated cyber attackers conduct their operations with little fear of identification. Attackers craft their operations such that even if the actor is identified it is unlikely, they will be apprehended and prosecuted in their country, or extradited to another country.

National security organizations have developed complex and effective infrastructures around the world using nearly untraceable financial accounts to create layers of complexity against detection and attribution. Criminal organizations have followed suit basing their operational decisions on the risk and likelihood of identification and arrest. Even unsophisticated attackers may use multiple hop points across varying jurisdictions to add extraordinary complexity to identification and attribution.

Criminal forensics technicians understand *Locard's Exchange Principle*; the perpetrator of a crime interacts with a crime scene leaving physical evidence at the crime scene and taking evidence from the crime scene, called trace evidence.[3] However, unlike the physical world cyber artifacts are digital information an attacker can create, erase, or manipulate to create false artifacts. In cyberspace there is often little physical evidence an investigator can point to and declare with absolute certainty.

Deliberate misdirection may benefit an attacker who can hide or leave disinformation to mislead investigators, discredit conflicting information, or support false conclusions. Disinformation may also assist a defender who can respond to remote attacks using purposely false information about the environment and its vulnerabilities to confuse attackers.

One of the earliest recorded hacker tracking operations used fake information embedded in data files to delay attackers while defenders traced the attackers back to their homes.[4]

Highly resourced organizations invest significantly into identifying those responsible for cyber intrusions. These organizations often believe these programs are sensitive, and knowledge of them must be protected from disclosure. If adversaries discovered the existence and capabilities of these attribution programs they could change their tactics to either cause attribution to fail, or exploit the attribution systems to misattribute adversary activities. This sensitivity for attribution system capabilities limits their use in criminal prosecutions or with media, and limits their use outside of sensitive government programs. This sensitivity effectively negates their deterrence potential.

Audacity

Cyberspace operations are characterized by a high degree of boldness. Attackers leverage anonymity and speed to conduct tactical events that may result in strategic effect. The low risk of attribution adds to the aggressive posture attackers can take in offensive operations. The Stuxnet worm allegedly targeted Iranian centrifuges in their nuclear program and epitomizes this boldness. The ability for an attacker to conduct audacious actions is underpinned by several components: Risk, Opportunity, Anonymity, and Jurisdictional Exploitation.

Risk. Every action that supports an operational or strategic goal carries risk. The attacker or defender must assess the potential gain against the likelihood of loss. Intelligence agencies refer to this process as "Intelligence Gain-Loss Assessment." Cyberspace operations typically carry a low risk to the attacker for many of the reasons already discussed.

Opportunity. The Internet has become the farmer's market of the world, having shrunk or eliminated borders for both legal and illegal actions and entities. Essentially all goods or services can be bought or sold on the Internet. It has also become an instantaneous storage device for billions of users. Cyberspace has also become a highway for gaining access to desired targets. The potential value of successful operations, ease of access, cost of access, anonymity, and limited threat of capture and prosecution provides tremendous incentive for audacious actions.

Anonymity. Security apparatuses around the world use highly trained operatives to conduct effective covert operation. Anonymity in these operations enables bold action. Likewise, boldness of action in cyber conflict takes full advantage of the element of anonymity.

Jurisdiction Exploitation. Many people and organizations understand the complications for law enforcement and intelligence agencies in gathering sufficient evidence to act. In the 2013 Annual Report to Congress, the United States Department of Defense implicated the Chinese Government in breaches of multiple governmental and commercial entities in the United States.[5] The attacks in this report are likely only a portion of activity the Chinese Government perpetrated against the United States Government and commercial entities. Despite the volumes of investigatory work and information collected government agencies have taken little effective action against individuals in China suspected of the attacks. Likewise, the United States indicted twelve members of the Russian GRU military intelligence for compromising the 2016 US election infrastructure and sowing discord among the American people. As noted earlier, many other countries are conducting similar operations in cyberspace, including the United States. More importantly, numerous international laws, including the 1907 Hague Land Warfare Regulations and the third and fourth Geneva Conventions of

1949, made international espionage legal.[6] It's unlikely any of these international defendants will be tried.

The characteristics of cyberspace fully empower actors who choose to take audacious actions. Regardless of whether they are attackers or defenders, audacity is a key principle in the perpetual conflict in cyberspace. While the meek may inherit the earth, it is the bold who will seize their place in cyberspace.

Initiative

Defense is generally accepted as the stronger form of kinetic combat. A defender can understand and exploit the battlespace to their advantage since they typically choose the terrain to defend. The defender can build fortifications to protect their forces, while delaying an adversary and conserving their military force. The defender doesn't have to defeat an attacker, only exhaust their opponent or convince them exhaustion will occur prior to the attacker achieving success. The defense may also give the defender a moral and political advantage by being viewed as the victim of an aggressor.

However, defense is passive, allowing an attacker to choose the time and place for operations. In kinetic operations commanders use offensive maneuvers in the attack and counter-attacks to seize, maintain, and exploit the initiative and instill their will upon the enemy. Offense enables an attacker to fight on their terms and risk their forces when they believe the conditions are favorable to them. This initiative allows the attacker to use their advantages against the enemies' weaknesses.

Eastern military thinking, driven primarily by Chinese military strategist Sun Tzu, focuses on initiative not offense. Sun Tzu understood the strength of the defense, but also the limitation, noting, "Invincibility lies in the defense, the possibility of victory lies in the attack." Sun Tzu embraced the concept that the most successful battle is one where the leader achieves their goals through alternative means without risking the loss of their forces.[7]

Cyberspace also favors the entity that seizes the initiative. This principle is one that can equally be shared by both attackers or defenders.

An attacker may exploit a vulnerability before the defender can either eliminate or mitigate the vulnerability or respond effectively to the attack.

The defender can also leverage initiative by maintaining a continually adjusting defensive posture. The adjustments may include updating software to remove vulnerabilities, installing new signatures to repel known attack vectors, and rapidly responding to alerts.

At a more granular level, the initiative in gained and maintained by leveraging persistence, patience, anonymity, surprise, and speed.

Persistence. Unlike traditional domains of warfare, cyberspace is omnipresent and is under attack on a daily basis. Threats can launch attacks with varying levels of finesse and intensity, depending on the temporal needs of their mission or operational focus. If their targets rarely change an attacker can methodically plan and execute operations using time to their advantage to continually 'pound' on targets until successful.

Patience. This element may seem out of place when considering Initiative. However, contrary to physical forms of warfare, patience is a key component of Initiative. As attackers learn about their target's environment, they may find they must wait for the right opportunity to present itself. For example, once an attacker gains access to an individual workstation they may need to remain on the system quietly awaiting an administrator to log onto the system for routine maintenance. Once the administrator has logged on the attacker captures the administrator's credentials and uses them to launch the next phase of their operation, now with the administrator's credentials and privileges.

Anonymity. Attackers leverage anonymity to gain and maintain the initiative. It supports the components of persistence and patience, and amplifies the element of surprise, enabling the attacker to conduct their operations with extraordinarily low risk of discovery and identification.

Surprise. In traditional warfare surprise occurs when one adversary takes an unanticipated action. This could be moving forces into a different area of operations or using deception. An adversary in cyberspace uses surprise in much the same way. Zero-day exploits and honey pots are examples of surprise.

Speed. In many fields operating quicker than an opponent may provide a significant advantage, to include getting inside an opponent's decision cycle.[8] The ability to send messages, commands, and malware across cyberspace from anywhere in the world and have automated systems respond in milliseconds makes speed-of-action much more significant.

The agility that attackers and defenders can apply to cyber actions is a significant difference compared to kinetic operations. An attacker can plan and launch a devastating attack in hours or days. In December 2010 the hacker consortium Anonymous launched flooding attacks against financial institutions in retaliation for denying transactions supporting Julian Assange of Wikileaks fame. Anonymous developed and launched this attack, which surprised authorities in its size and effect, and successfully prevented customers from accessing Visa and Mastercard websites.[9]

Asymmetric Effects

In many fields including kinetic warfare, an effect is proportional to the effort and resources applied. During the 2014 Crimean independence referendum, Russia imposed its will on the land and air due to the resources it possessed in that region of the world. No nation was willing to contest Russia on the Crimean Peninsula in either of those domains of war. In this regard, there is symmetry to the resources applied when compared with the anticipated results.

In cyberspace the effect of an offensive or exploitative event can be greatly disproportional to the resources applied. Indeed, many cyber breaches today are initiated by a simple phishing or watering hole attack. The resources required to craft and send a phishing email and packaging an exploit of a known, yet unpatched vulnerability is low. Yet the benefits of a successful large-scale breach to the attacker may include terabytes of sensitive data or significant financial gains. Additionally, threats can employ deceptive tactics such as a *false flag* operation. Where they deliberately conduct actions to appear as one person or organization, when they are not. *False flag* operations can fool defenders so that they do not look deeper into nefarious actions, or they could be overt in a way that draws resources toward a deception operation while enabling the attack's main effort to go unnoticed.

An uninformed or poor decision by a single person could result in an attacker compromising an entire corporate network, and exfiltrating the organization's entire personnel, financial, or national security information. Alternatively, the attacker could use their access to change information in databases, order supplies or change existing orders, misdeliver or cancel orders, or misdirect national satellites. Attacks against information systems can affect physical environments, like deactivating river locks, opening dam floodgates, deactivating networked electric power control systems, or damage or control of seagoing vessels.

In security fields it's often touted that defenders can't defend against a "stupid user." However, many successful phishing or waterhole attacks use previously identified vulnerabilities that would not have been successful if security personnel had maintained their systems. Also, many attacks leverage the normal functionality of a network that could have been protected by reasonable security practices. In these cases, leaders approved placing critical information and systems online when a simple action by a "regular" user results in catastrophic damage that defeats a reasonable defense.

The asymmetry within cyberspace exists primarily due to the imbalance between the impacts that can be inflicted against the effort to achieve the goal. The result is that countries with small or less capable resources can potentially inflict major damage to another country through the use of cyberspace.

Dependencies. Robert Metcalfe pointed out that there are exponential gains that the users of a network can obtain by connecting more users, and their information and systems. Antagonists quickly learned the converse of Metcalfe's Law is removal or isolation of nodes from a network reduces the value of the network by the square of the number of isolated nodes.[10]

The synergies we gain from interconnecting systems adds to our reliance, and the potential targets offered attackers. For example, an attacker could target credit card systems, shipping services, power systems, or Internet routing systems to significantly degrade on online book seller, without touching the book seller itself. An attacker can also pepper social media with false allegations about the bookstore that have to be refuted in public. Today critical national infrastructures, government systems, and military power projection are similarly composed of mutually interdependent systems-of-systems, and are therefore similarly vulnerable through their suppliers.

Exploitable vulnerabilities. Although banks get robbed occasionally, most thieves do not target banks. This is due to the belief that banks have eliminated existing vulnerabilities in their physical security, thus making a robbery a high-risk endeavor. If a bank was known to have flaws in its security the likelihood it would be targeted for theft increases markedly. The sheer volume of software that exists today and its corresponding vulnerabilities creates an environment that begs to be targeted by various threats around the world.

Reproducibility. On a kinetic battlefield there is conservation of weaponry. Some weapons, such as bombs and missiles are expended and destroyed when they are deployed. Other weapons, such as knifes are used repeatedly but are degraded over time due to use. Still other weapons, such as modern assault rifles, can be used repeatedly, but are limited to the quantity of expendable ammunition a service member can carry.

Cyber weapons do not suffer from these restrictions. An individual or group can make unlimited copies of a cyber weapon or use a cyber weapon repeatedly at no additional cost.

This ability to record, store, share, and reuse digital weapons also enables the victim of an attack to capture a weapon and launch it on the perceived perpetrator or anyone else. Some believe Iran perpetuated the destructive cyber-attack against the Aramco corporate enterprise in retaliation for two previous U.S. cyber-attacks and reused the code from one of these attacks.[11]

Since computer programs may make copies of code and data they can also make unlimited copies of themselves. A computer worn does just that, by making copies of itself and sending them to other computers. Computer worms could create a digital pandemic of infected machines. Infected machines can also degrade other systems even if they don't infect the machines. The 2001 Code Red worm degraded network services when it flooded networks with copies of itself. These malicious computer programs may also change their code while retaining their original functionality. This *polymorphic* property can further hinder detection since retrospective defenses may not identify mutated version of malware.

Some defense contractors and vendors have promoted the *Cyber Kill Chain* to describe the cyber attacker's process.[12] The Cyber Kill Chain illustrates the artifacts defenders may observe as a perpetrator steps through an attack. However, researchers have documented serious flaws with the process.[13] Among these is that the process does not recognize that sophisticated, resourced attackers develop infrastructures to automate their preparation, execution, and recovery operations. These infrastructures make an attacker resilient in the face of a dynamic defender. Rather than respond to individual cyber-attacks the security community could attack a malicious group's infrastructure. Degrading and denying attacker's infrastructure could create asymmetric effects against the attacker as they consume time and resources to rebuild infrastructures and not conduct offensive operations.

Economy of Force

The economics of planning and executing full-scale warfare on the air, land, sea or space can have a significant impact on the national treasure of a belligerent country. Even a limited incursion has significant monetary, political, and societal costs. Defending against kinetic aggression can be equally costly. Similarly, planning a multi-million-dollar heist requires extensive planning and resources that may come at considerable cost and risk to a criminal.

Planning and executing full-scale offensive operations in cyberspace can be orders of magnitude less costly than physical space. This principle is recognized and leveraged by states, transnational entities, criminals, and others as they conduct operations in pursuit of their goals. This principle contains three central elements: Cost of Entrance, Cost of Operations, and Collateral Damage.

Cost of Entrance. Within the physical domains of warfare, the ability to project firepower across an ocean or a major landmass is a tremendous undertaking of national resolve and national treasure. Alternatively, a cyber-attacker can exploit the relatively low cost of computing technology to create effective cyber weapons to project power worldwide in seconds, and at a fraction of the cost.

Cost of Operations. Once armed with sufficient tools and personnel to conduct offensive operations in cyberspace, the cost to continue operations pales compared to air, land, sea, and space operations. The U.S. Army and U.S. Marine Corps have many tens of thousands of infantry personnel within their ranks. Even with all of these individuals the United States National Military Strategy states the United States could only conduct one major and one minor war simultaneously. Effective cyber operations can be conducted by a quantity of personnel orders of magnitude less than physical domains. The hactivist group Anonymous has conducted multiple and extensive operations using volunteers who simultaneously partook in a wide variety of cyber activities across the globe. Attackers can develop and execute operations using technologies to enable a single actor to launch 'fronts' against multiple targets. The cost of continued operations is so low attackers can rent 10,000 compromised computers per day to aid in their operations for as little as $1,000 per day.[14]

<u>Collateral Damage.</u> Collateral damage can be a defining data-point when considering if a kinetic action should go forward. A leader may cancel or postpone a planned strike against a hostile target due to the target's proximity to a place of worship, hospital, or a school. If collateral damage may occur in an offensive cyber-attack, the attacker can use the deniability afforded by multiple layers of anonymity into their calculus for their decision to conduct the operation. If a carefully planned and executed cyber-attack offers total anonymity a leader may not care whether collateral damage occurs, since investigators may not trace the activity back to them. Stuxnet, agent.btz, and the Target Corporation breach are prime examples.

Cyber operations, like kinetic operations, may create external costs to and from parties not directly targeted by an attack. For example, a cyber-attack may require the attacker transmit malware or command codes through many intermediary systems to the victim. Any of these parties may record the code and reuse it against additional victims. Thus, when an adversary uses a new vulnerability or weapon it is considered "burned" and available to others.

A cyber attacker also has direct costs for planning and executing operations in cyberspace. Successful and highly targeted malware requires time-consuming and customized design to operate on specific systems. These attack tools are frequently brittle and fail to degrade or disable a system due to any change to the targeted system. A system patch or update may remove the vulnerability or other system functionality the malware relied upon, or security tools may detect the malware. In either case, the attacker's mission may fail and the defender may detect the attacker and capture the weapon.

Likewise, governments may spend tremendous resources discovering and weaponizing vulnerabilities inside costly "classified" programs. Governments use secrecy to ensure the cyber weapons will be effective when used. However, defenders are also searching for vulnerabilities. If defenders independently discover a vulnerability they may inadvertently "burn" a secret weapon prior to its use, negating its use and the expense that went to discover and weaponize the vulnerability. In just one three-day security conference hackers discovered 51 zero-day bugs.[15] These bugs were shared with vendors to improve the security of their systems, and thus eliminate their use by attackers.

There are few economies of force for the defender. While a defender may install and maintain numerous security controls,[16] there are few studies on the effectiveness of the controls, the effect of composing controls together, impact to operations of the controls, cost to install and maintain the controls, or new vulnerabilities introduced by the controls as they are installed and maintained over time.

Resilience

The unique nature of cyberspace is a significant contributing factor in the resilience of cyber operations. Most Internet residents have accidentally deleted files, been affected by a virus, inadvertently installed unwanted programs, or received the 'blue screen of death.' If an Internet resident does not possess the skills to recover from these events they can contact their friend, relative, or co-worker to help them restore their systems and information.

Many governmental and commercial organizations have heavily invested in their networks to ensure they are resilient in the face of threats. They may have established a comprehensive incident response function, created information back-up processes, established a business continuity capability, or implemented a disaster recovery site.

Cyberspace can offer similar resilience. In December of 2015, hackers attacked a portion of the Ukrainian power grid and denied power to approximately 225,000 customers.[17] Despite the effectiveness of the attack, power was restored and systems were functioning within two days. This outage demonstrated the vulnerability of industrial control systems and the resilience of motivated defenders and operators to quickly restore systems.

Likewise, in March 2019 journalists reported the U.S. Cyber Command, the organization responsible for attacking foreign systems and defending U.S. military system, attacked foreign civilian targets. They claimed the US military disrupted hackers' systems as they were manipulating US elections through deceptive social network messaging. While "unnamed government officials" claimed the government attack was a success, it most likely served only as an annoyance until the hackers could restore their systems.[18]

It's important to realize most attacks, if successful at all, will create temporary effects to the victim, and announce knowledge of the vulnerability to the victim. A rational victim may then remove the vulnerability from their systems. Thus, after recovery the defender is stronger.

Attacks that exfiltrate sensitive information, such as the 2014-2015 attack on the Office of Personnel Management, may create lasting effects. It's believed that during this attack foreign hackers captured the personnel and security records of 21.5 million former and current US government employees. FBI Director James Comey stated: "It is a very big deal from a national security perspective and from a counterintelligence perspective. It's a treasure trove of information about everybody who has worked for, tried to work for, or works for the United States government."[19] Foreign governments could use this information to advance their goals.

Heterogeneity

In the 1980's the United States military transitioned from the .45 caliber pistol to the 9mm pistol as its primary sidearm to ensure interoperability with European weapons and ammunition stockpiles. The Soviet Union and Warsaw Pact applied this same concept across their weapons and munitions. In the physical domains of warfare, homogeneity gives organizations greater efficiencies, with uniform processes and effectiveness. This may result in significantly reduced costs. However, adversaries may exploit this efficiency and effectiveness.

<u>Offense.</u> Organizations around the world have widely varying hardware, firmware, and software configurations, and varying competencies in applying and maintaining security controls. These differences necessitate attackers leverage different attack

platforms and packages to effectively conduct their operations. They may leverage a *botnet* to conduct distributed reconnaissance, utilize a compromised Linux system to launch an initial attack from one country, and establish command and control in the target environment with a Windows-based system from an entirely different country. Each of these operations may employ different software packages, different platforms, and different national jurisdictions. This heterogeneity may allow for a greater probability of success while simultaneously reducing the chance of compromise of the entire operation and the attacker's support infrastructure.

Defense. Many organizations place the same family of anti-virus software on their workstations and servers. In this situation an attacker needs only defeat or bypass one type of security control to compromise both workstations and servers. Heterogeneity in the defense may enable an organization to identify and mitigate more attacks. Multiple versions of security tools applied throughout the environment increases the organization's chance of detecting malware. Likewise, an organization with multiple Internet connections may apply firewalls from different vendors on their various gateways. This allows the defender to reap the potential benefits of different technologies and signature sets to monitor its assets.

A homogeneous environment may result in an optimal system design if the probability of exploitation is low. A common environment is easier to maintain and may provide economies of scale. However, as the probability of exploitation increases the probability of catastrophic "one-and-done" attacks increase. That is, exploiting one system may enable an attacker to jump to other systems, creating a system-wide cyber epidemic. Thus, heterogeneity may be a more effective in today's porous cyber environments, especially in mission-critical situations.

Dispersion of Effort

Traditional kinetic operations concentrate combat power against an enemy at a critical location and time. Massing forces in cyberspace operations, which often requires stealth, brings a greater chance of detection. This creates an inverse principle for cyber operations, the dispersion of effort.

An attacker can disperse efforts by using various locations to conduct multiple elements of their operation. This reduces the attacker's risk that if a defender detects one component of an attack, they will detect the other components. Another aspect of dispersion, is segmenting work functions. Attackers can divide the effort among teams based on their skills. A sophisticated group may consist of a Team Leader (establishes the operational goals), code-writer (exploit developer), infrastructure engineer (establishes and maintains obfuscated command and control network), social engineering specialist (phishing email creator), network specialist (expands access in the target environment), deception (conducts operations to distract the target from the main operation), and interpreter (translates captured documents). In much the same way intelligence and terrorist organizations compartmentalize cells, a hacker team can be dispersed across many countries to complicate investigations, since team members may not know the existence or identity of the other members.

Under most circumstances, defenders are physically or logically collocated in cyberspace and must advertise their static location to accomplish their tasks. This provides attackers with a significant advantage in collecting intelligence on the defender's organization, technology, and people. Homogeneous environments (such as with enterprise site licenses) and bureaucracies that delay rapid prevention and response processes may further aid an attackers' reconnaissance and attack efforts.

It may appear an exception to the principle of Dispersion of Effort is denials of service (DoS) and distributed denials of service (DDoS) flooding attacks. Groups like the Syrian Electronic Army and Anonymous have massed thousands of computers to send messages to designated targets to degrade services. However, upon closer inspection these attacks use computers dispersed across the globe to mass attacks against concentrated systems. This further emphasizes that dispersion, not concentrating forces, may be a better principle for cyber operations.

Chapter 4. Conclusion

Conflict in cyberspace is a complex concept. Leaders who influence laws, policies, and budgets must spend the time necessary to grasp the properties of cyberspace. There are distinct and stark differences between the threats from physical (or kinetic) adversaries, and the threats faced by adversaries in cyberspace. Leaders must understand the concepts that govern the actions of the threats, the technologies and means used to wage cyber-conflict. Second, they must understand the various goals of both docile users and malicious actors. Third, leaders must understand the principles that govern conflict in cyberspace. They must learn how these principles apply to their private citizens, commercial entities, and national issues and how they may be leveraged by the various threats.

> Principle of Conflict in Cyberspace:
> - Anonymity – Reduce your costs by concealing your identity. Increase opponents' costs by identifying them.
> - Audacity – Exploit the low risk and high opportunities offered by cyberspace.
> - Initiative – Seize and retain the ability to instill your will upon opponents.
> - Asymmetric Effects – Exploit the battlespace to generate disproportional effects.
> - Economics of Force – Conserve resources.
> - Resilience – Design to absorb and recover from effects when planning and executing operations.
> - Heterogeneity – Diversity may be more effective in highly connected and vulnerable systems.
> - Dispersion of Effort – Distribute resources across the battlespace to reduce risks.

Conclusion

The Principles of Conflict in Cyberspace are the start of a journey to frame a discussion imperative to the future welfare of the United States and its allies. However, like the Principles of War, these principles are neither absolutes nor checklists for success. Government leaders must weigh each principle against the situation where they find themselves.

Our understanding of the cyberspace domain and principles that govern operations in the domain, should drive how we operate in cyberspace. For example, a nation could retain knowledge of an unpublished vulnerability that may provide them with an indefensible attack vector. However, delaying the publication of this zero-day, while holding the vulnerability in reserve and not using the vulnerability, violates the Principle of Initiative. Other parties (e.g. a hostile nation's intelligence service) could independently discover the vulnerability and seize the initiative by publishing or using the vulnerability. This doctrinal question of "what to do with zero days" should be analyzed and debated within the construct of Principles of Cyber Conflict.

As a case study, the Stuxnet virus may have provided a nation or nations with a tactical win, but may have resulted in numerous large-scale losses. The apparent goal of the malware was to delay the Iranian nuclear effort. The costs included providing the world with knowledge of numerous vulnerabilities embedded in the software that others used in later attacks including the August 2012 destruction of 30,000 Aramco oil company computers. Some believe this attack was a deliberate retribution attack by Iranian agents in response to Stuxnet.

The Stuxnet malware proved that attackers could use malware to cause destructive physical attacks, serving to motivate others to research similar attacks. Lastly, Iran has used their perceived victimization by the Stuxnet malware as justification for conducting attacks against the United Stated and its allies, motivating nations to further develop and refine attacks, and possibly leading to increased willingness to further weaponize cyberspace.

Conflict in cyberspace has an elastic quality. To a degree, a version of Sir Isaac Newton's Third Law of Motion[1] is at play; *'for every action there is an equal and opposite reaction*. The harder an entity presses on other cyber residents, the harder they push back, and this perpetual backwards and forwards nature will likely have no ultimate winners. When one organization appears to obtain a disproportionate level of influence or control they draw the attention of other cyberspace residents who may wish to level the playing field. This leads us to believe that macro level conflicts in cyberspace are more likely a no-win endeavor for all participants.

Additionally, cyberspace and cyber conflict have an evolutionary quality. These principles exist for a particular environment. A significant technological change, such as quantum computing or true artificial intelligence systems, could fundamentally change these principles.

Simply put, there may not be a means to "win" in cyberspace. Just as it may be impossible to eradicate crime or warfare, it may be impossible to eradicate cyber-conflict. However, nations need to develop a long-term strategy to protect their interests, if leaders understand the environment in which they operate, and the implications of their decisions.

Glossary

Administrator Access — A privileged account permitting a higher level of authority and unrestricted freedom in what the account holder can perform on systems.

Active Defense — The employment of limited offensive actions and counterattacks to temporarily disable an imminent and impending attack.

Advanced Persistent Threat — Hackers who establish a stealthy and continuous presence on a victim's network using sophisticated capabilities and/or tradecraft.

Asymmetric — An out of balance or unequal situation. An asymmetric advantage is an advantage providing a much greater benefit then the costs to achieve the benefit.

Asymmetric Warfare — A conflict where the resources of two combatants differ and each attempt to exploit the other's weaknesses.

Attack Surface — The collection of all vulnerabilities a hacker may use to achieve their goals.

Attack Vector — A specific means to deliver malware or commands to a target.

Attribution — The ability to definitively identify and assign responsibility for any actions to a machine or individual on a network.

Backdoor — Any method permitting the bypass of normal access controls in a system.

Beacon — Software repeatedly transmitting messages indicating it is active. Hackers install beacons to call out of compromised hosts to the hacker, who can then accept the connection and bypass security controls.

Best Security Practices — A list of security processes that work in the general case. Examples include knowing what is installed on a network, limiting control and access to information and systems, maintaining systems with the latest patches and updates, monitoring everything, and having a practiced response plan.

Botnet — A set of compromised (zombie) computers controlled by a master.

Bug — an error in computer code.

Cloud Computing — Outsourcing information services.

Collateral Damage — Unintentional or incidental injury or damage to persons or objects that would not be lawful military targets in the circumstances ruling at the time. (US Department of Defense Joint Publication 3-60)

Controls — Any manual or automated (security) tool meant to ensure the quality of a state or process.

Critical Infrastructures — "The systems and assets, whether physical or virtual, so vital to the United States that the incapacity or destruction of such systems and assets would have a debilitation impact on security, national economic security, national public health or safety, or any combination of those matters." USA Patriot Act of 2001.

Cut-Out/Hop-Point — Compromised third-party computers that are unwitting intermediaries between a hacker and their victim. These are often used as hop-off points to attack a victim, or drop-off points to send exfiltrated information.

Cyberconflict — Human struggle and ambition projected onto cyberspace. It is the actions of belligerents in cyberspace where conflicting agendas have transformed information and information systems into the targets, weapons, and battlefield.

Cyberspace — The set of computers and people interconnected by communications systems to enable them to achieve their goals.

Default Passwords — The passwords that come installed on systems by the vendors and are easily found by hackers.

Defense-in-Depth — Redundant security controls placed in series in the hope one detects and/or prevents malicious activities.

Denial of Service (DOS) — An attempt to make a machine or network system unavailable by overwhelming its processing, storage, or network resources.

Egress Filtering — Removing potential malware or other malicious activity as it leaves a network.

False Flags Operations — Deliberately using untrue information to mislead victims of a perpetrator's identity. This phrase goes back to sailing ships, where a captain would fly the flag of another country to deceive another ship captain, then 'strike' that flag and raise their national flag when it was too late for their victim to react effectively.

Geolocation — Assigning a geographic location to a person, device, or activity.

Honeypot/Honeynet/Honeymonkey — A honeypot is data and/or systems that appear legitimate but that defenders use to bait and/or observe hackers. A honeynet is a set of honeypots. A honeymonkey is an automated system that visits websites searching for sites that infect visitors with malware.

Insider — Any member of a group (e.g. current or former employee or contractor) who has taken a hacker role such as voyeur, spy, or saboteur.

Intelligence — Activities to collect, analyze, and disseminate information so decision makers can grasp what is occurring or about to occur and make informed decisions.

Intelligence Community (IC) — In the U.S. it is a collection of sixteen government agencies that perform intelligence activities on behalf of the U.S. Government. The IC is led by the Director of National Intelligence (DNI).

Malware — Malicious software. Any software that a hacker may use to create a desired effect. See Trojan Horse, Virus, and Worm.

Mark — Intended victim.

Metadata — Data about data.

Obfuscate — To change a file or application format or structure, to make its functionality unclear or unintelligible.

One-and-Done — see Zero-Day Vulnerability

Open Source Intelligence — Collecting publicly available information.

Owned — Hacker slang for having complete control of a victim's computer.

Patch — Computer code that fixes an error or adds new functionality in an application or operating system.

Phishing — A method to acquire sensitive information by masquerading as a trustworthy entity in electronic communications.

Protocol — A set of rules for communicating.

Pyrrhic Victory — A military victory that is so costly to the "winning side" that they are worse off.

Resilient Systems and Networks — Strategies to quickly restore systems when they break or are compromised.

Retrospective Analysis — The progression from failure, detection, investigation, and update that most security controls require to defend against new attacks.

Reverse Engineering — Analyzing systems structure, function and operation of computer code and hardware to determine how a system operates.

Risk Management — A formal process to conduct informed guessing based on unknown vulnerabilities, adversaries' capabilities, effectiveness of security controls, or impact of an attack.

Sneaker Net — Moving data between networks using removable media.

Social Engineering — Conning, defrauding, and misleading.

Spam — Unsolicited or unwanted messages.

Spills — A euphemism for the loss of an organization's sensitive information.

Spoof — To imitate.

Spycraft — See Tradecraft.

Super User — See Administrator Access.

Superiority — A condition when a side has a more favorable position than their opponent in an environment. See Supremacy.

Supply Chain Attack — Compromising a system typically by entering malware or counterfeit components into products during the design, manufacturing, handling, distributing, and installation stages of the manufacturing process.

Tradecraft — The techniques and activities used to achieve a goal. In the Intelligence Community this term is associated with activities by human intelligence agents to enable the gathering of intelligence information, rather than other more technical intelligence collection.

Tragedy of the Commons — A situation where each individual's pursuit of their own best interest results in a situation that is worse off for the population.

Trojan Horse — Malware that appears to be a useful application or data but contains malicious content, similar to the wooden horse of Greek mythology. When the recipient accesses the file or program the Trojan payload is activated.

Trusted System — A system that has some security properties that causes decision-makers to believe their system is secure or safe against specific threats.

Virus — Malware attached to a legitimate program or file. When someone accesses a program or file infected by a virus the virus infects the computer with the virus payload commands. A virus spreads when a human or program shares the infected program or file with another machine.

Watermark — Unique sets of characters or images hidden in files.

Weaponization — The process of converting bugs and vulnerabilities into effective and efficient offensive tools.

Worm — Malware that uses a system's existing communications features to spread to other machines.

Zero-day Vulnerability — A vulnerability that has not been publicly announced, and therefore is most likely not defended against.

Zombie — A compromised computer controlled by a hacker. See Botnet.

References and Notes

Chapter 1

[1] Applying the Principles of War to cyberspace is a very popular subject. A search of the phrase "principles of cyber war" returned 975,000 pages on an Internet search engine. Documents reviewed include:

- Samuel Liles, J. Eric Dietz, Marcus Rogers, Dean Larson. *Applying Traditional Military Principles to Cyber Warfare*, 4th International Conference on Cyber Conflict. C. Czosseck, R. Ottis, K. Ziolkowski (Eds.). 2012, Tallinn downloaded from http://www.ccdcoe.org/publications/2012proceedings/3_2_Liles&Dietz&Rogers&Larson_ApplyingTraditionalMilitaryPrinciplesToCyberWarfare.pdf on 21 April 2014.
- Steven E. Cahanin. *Principles of War for Cyberspace*. Air War College Air University. 15 January 2011 downloaded from http://www.nsci-va.org/CyberReferenceLib/2011-01-15-Principles%20of%20War%20for%20Cyberspace%20Research-Cahanin.pdf on 21 April 2014.
- David B Farmer. Do the Principles of War Apply to Cyber War? School of Advanced Military Studies United States Army Command and General Staff College Fort Leavenworth, Kansas AY 2010 downloaded from http://handle.dtic.mil/100.2/ADA522972 on 21 April 2014.

[2] In kinetic warfare state and non-state actors may view citizens as protected entities, distractions, or a source of labor and resources, as well as provocateurs, spies, or guerillas.

[3] "50 U.S.C. § 413b(e)." Title 50, United States Code, Section 413b, *Presidential approval and reporting of covert actions*.

Chapter 2

[1] Moore's Law states the number of transistors in an integrated circuit doubles approximately every two years. The capability of many computing devices is linked to the number of transistors.

[2] Kryder's Law states a hard drive density increases by a factor of 1,000 every 10.5 years.

[3] The value of a network, and Robert Metcalfe's formula, are contested subjects. It may be more accurate to say as more systems and users (n) are added to a network the relative benefit people receive from the system or network increases by some value between $n \log n$ and n^2.

[4] http://www.internetworldstats.com/stats.html.

[5] Stoll, C., *The Cuckoo's Egg – Tracking a Spy Through the Maze of Computer Espionage*, Pocket Books, 1990.

[6] Federal Bureau of Investigation. *2017 Internet Crime Report.* Accessed on June 1, 2018 at http://www.ic3.gov/media/annualreport

[7] https://www.symantec.com/content/dam/symantec/docs/about/2017-ncsir-global-results-en.pdf

[8] The author does not wish to imply that cyberspace is another "domain" per the U.S. DoD philosophy. The DoD has not defined 'domain' thus the concept of a "cyber domain" may be a self-fulfilling prophecy to achieve their political goals.

[9] *DoD Strategy for Operating in Cyberspace*, July 2011, http://www.defense.gov/news/d20110714cyber.pdf

[10] U.S. Navy Rear Admiral Grace Murray Hopper is credited with popularizing the term 'computer bug' after a moth caused her computer to malfunction.

[11] McConnell, Steve. *Code Complete, Second Edition.* Microsoft Press. 2004. pp 521-522. This text provides an excellent technical discussion on the source of errors in computer code and how to eliminate them.

[12] These numbers are interpolations based on estimates extracted from informed experts and empirical data.

[13] NISTIR 7298, Revision 2, *Glossary of Key Information Security Terms*, May 2013.

[14] There are other threats to information systems. These include natural and physical threats such as fires, floods, power failure, smoking and soda spills, that can be reduced through policies, preventive actions and damage minimization; and unintentional threats, due to ignorance and accidents (such as erasure of data or incorrect data entry). Russell, D., and Gangemi, G.T., *Computer Security Basics.* O'Reilly & Associates, Sebastopol, CA, 1991.

[15] Frederick Brooks noted, "All repairs [to software] tend to destroy the structure, to increase the entropy and disorder in systems." Frederick Brooks, *The Mythical Man-Month*, (P122), https://archive.org/details/mythicalmanmonth00fred

[16] A discussion of undecidability is beyond the scope of this paper. In layman's terms; it has been mathematically proven impossible to create a general-purpose computer that can take as input any computer code, and tell if it is malicious. This makes it impossible to build a general-purpose anti-virus, intrusion detection system, or software bug detector. See http://all.net/books/Dissertation.pdf

[17] This is not to say profit motive is a bad thing, simply it is a key motivation today.

[18] Ken Thompson. "Reflections on Trusting Trust." *Communication of the ACM*, Vol. 27, No. 8, August 1984, pp. 761-763. Accessed May 6, 2014 from http://cm.bell-labs.com/who/ken/trust.html

[19] Attacks have already been reported in power meters (http://www.greentechmedia.com/articles/read/hack-your-meter-while-you-can), refrigerators (http://www.cnbc.com/id/101434941), and pacemakers (http://www.forbes.com/sites/singularity/2012/12/06/yes-you-can-hack-a-pacemaker-and-other-medical-devices-too/).

[20] https://www.intel.com/content/www/us/en/architecture-and-technology/side-channel-variants-3a-4.html

[21] https://www.intel.com/content/www/us/en/architecture-and-technology/side-channel-variants-1-2-3.html

[22] *Government-sponsored cyberattacks on the rise, McAfee says.* Networkworld.com (29 November 2007). Retrieved from http://www.networkworld.com/news/2007/112907-government-cyberattacks.html on 22 April 2014.

[23] DoD Dictionary of Military and Associated Terms. Joint Publication 1-02. 8 November 2010, (As Amended Through 15 March 2014)

[24] http://threatpost.com/groundbreaking-cyber-fast-track-research-program-ending-030613/77594

[25] http://www.pcworld.com/article/2150743/antivirus-is-dead-says-maker-of-norton-antivirus.html

[26] NIST Special Publication 800-53A provides a process to ensure security controls meet their intended goals. For example, for

"WIRELESS ACCESS" the document tells managers to "Determine if the information system protects wireless access to the system using authentication and encryption." Unfortunately, there isn't guidance to managers as to the cost of implementing authentication and encryption, or if the specific technologies deployed to meet this objective are effective at deterring or preventing attacks.

[27] S. Frei, *Correlation of Detection Failures*, NSS Labs, 2013. Retrieved from https://www.nsslabs.com/news/press-releases/are-security-professionals-overconfident-%E2%80%9Cdefense-depth%E2%80%9D on 1 February 2014.

[28] Xue reports on 165 vulnerabilities over a four-year period in anti-virus technologies, concluding "[It] is clear that antivirus software can be targeted just likes other components or services of computer systems." Feng Xue, *Attacking Antivirus*, retrieved from http://sebug.net/paper/Meeting-Documents/syscanhk/AttackingAV_BHEU08_WP.pdf on May 20, 2014.
The Heartbleed exploit uses a hole in Secure Socket Layer tools designed to allow secure information sharing on the Internet. http://www.cnn.com/2014/04/08/tech/web/heartbleed-openssl/

[29] President's Information Technology Advisory Committee. *Report to the President -- Cyber Security: A Crisis of Prioritization*. Feb 2005. Edited by Marc R. Benioef, Edward D. Lazowska.

[30] Lolita C. Baldor, *Cyber Weaknesses Should Deter US from Waging War,* Associated Press, November 8, 2011, retrieved from http://news.yahoo.com/cyber-weaknesses-deter-us-waging-war-212947418.html on April 22, 2014.

[31] Douglas W. Hubbard. *The Failure of Risk Management: Why It's Broken and How to Fix It*. Wiley Publishing, 2009.

[32] Richards J. Heuer, Jr, *Psychology of Intelligence Analysis*. Available at https://www.cia.gov/library/center-for-the-study-of-intelligence/csi-publications/books-and-monographs/psychology-of-intelligence-analysis/

[33] See http://www.l0phtcrack.com/Beautiful_Security_chapt_1.pdf

[34] Attribution is determining the identity or location of the person or organization responsible for an attack.

[35] The Ponemon Institute, *2018 Cost of a Data Breach Study: Global Overview*, July 2018

[36] https://ieeexplore.ieee.org/document/1391513/
[37] Metadata is data about data.
[38] http://kieranhealy.org/blog/archives/2013/06/09/using-metadata-to-find-paul-revere/
[39] Visa. *Online Fraud Benchmark Report: Persistence is Critical.* 2017. https://www.cybersource.com/content/dam/cybersource/2017_Fraud_Benchmark_Report.pdf
[40] Being compromised is admittedly more serious for organizations advertising themselves as cyber security or national security specialties, such as government contractors, security vendors, or government security agencies.
[41] U.S. Federal Trade Commission. The Equifax Data Breach. https://www.ftc.gov/equifax-data-breach
[42] Kim Zetta, *NSA Hacker Chief Explains How to Keep Him out of Your System.* Wired Magazine. January 28, 2016.
[43] On September 26, 2018 – Google.com returned 43,200,000 results to this query.
[44] Compiling a computer program, or executable, is the process of converting human-readable computer instructions (called source code) to computer-readable computer instructions (sometimes called binary or object code).
[45] https://www.backtrack-linux.org/
[46] https://www.justice.gov/usao-ndga/pr/two-major-international-hackers-who-developed-spyeye-malware-get-over-24-years-combined
[47] W. Arbaugh, W. Fithen, J McHugh., *Windows of Vulnerability A Case Study Analysis.* IEEE Computer, December 2000.
[48] VanPutte, Michael A. *Walking Wounded – Inside the US Cyberwar Machine.* Createspace. 2016.
[49] Note: Making a false report to an emergency or law enforcement agency is a felony in nearly every state and municipality. A hacker was sentenced to eleven years in prison for his swatting activities. See http://www.wired.com/2009/06/blind_hacker/

Chapter 3

[1] William J. Lynn III, Foreign Affairs, *Defending a New Domain, The Pentagon's Cyberstrategy*, September/October 2010. Retrieved June 1, 2012 from http://www.foreignaffairs.com/articles/66552/william-j-lynn-iii/defending-a-new-domain

[2] The authors admit they cannot find a military definition of *domain*, and without this definition Mr. Lynn's declaration appears a self-fulfilling prophecy.

[3] Paul Leland Kirk. *Crime Investigation: Physical Evidence and the Police Laboratory.* Interscience Publishers, New York. 1953.

[4] See Cliff Stoll, *The Cuckoo's Egg.*

[5] http://www.defense.gov/pubs/2013_china_report_final.pdf

[6] http://www.faqs.org/espionage/In-Int/Intelligence-and-International-Law.html#ixzz339EenGmW

[7] Sun Tzu (Author), Thomas Cleary (Translator). *The Art of War.* Mass Market Paperback. 2005.

[8] Military strategist and USAF Colonel John Boyd is often credited with developing the concept of the OODA loop, for observe, orient, decide, and act. Boyd never published a paper on the OODA loop, however a five-slide presentation credited to Boyd titled *The Essence of Winning and Losing*, can be found at http://www.danford.net:80/boyd/essence.htm

[9] Addley, Esther; Halliday, Josh. *WikiLeaks supporters disrupt Visa and MasterCard sites in 'Operation Payback* December 09, 2010. (London:The Guardian). Retrieved 2010-12-09 from http://guardian.co.uk.

[10] A simple example using an arbitrary utility value is a network of 500 nodes would generate 500^2, or 250,000 utility. If a hacker disables a routing device and divides the network into independent 200 and 300 node networks, the total utility is now $(200^2) + (300^2)$ or 130,000 utility, effectively reducing the benefit users receive from the network by nearly half.

[11] http://www.nytimes.com/2012/10/24/business/global/cyberattack-on-saudi-oil-firm-disquiets-us.html?pagewanted=all&_r=0

[12] E. Hutchins, M. Clopperty, R. Amin. *Intelligence-Driven Computer Network Defense, Informed by Analysis of Adversary Campaigns and Intrusion Kill Chains.* Lockheed Martin Corporation. Available from http://www.lockheedmartin.com/content/dam/lockheed/data/corporate/documents/LM-White-Paper-Intel-Driven-Defense.pdf

[13] VanPutte, M. A. *Tracking, Degrading, and/or Disabling Cyber Adversary Support Infrastructures.* DARPA Cyber Scope Project. Unpublished DARPA Research Paper. February 2016.

[14] https://threatpost.com/how-much-does-botnet-cost-022813/77573

[15] Zero-Day Initiative. *The Results – PWN2OWN 2017 Day Three.* https://www.thezdi.com/blog/2017/3/17/the-results-pwn2own-2017-day-three

[16] See NIST Special Publication 800-53, Security and Privacy Controls for Federal Information Systems and Organizations, DoD Risk Management Framework, and Intelligence Community Directive 800-53.

[17] http://www.zdnet.com/article/us-report-confirms-ukraine-power-outage-caused-by-cyberattack/

[18] Ellen Nakashima. "U.S. Cyber Command operation disrupted Internet access of Russian troll factory on day of 2018 midterms." *The Washington Post.* February 27, 2019. https://www.washingtonpost.com/world/national-security/us-cyber-command-operation-disrupted-internet-access-of-russian-troll-factory-on-day-of-2018-midterms/2019/02/26/1827fc9e-36d6-11e9-af5b-b51b7ff322e9_story.html?noredirect=on&utm_term=.0c5298dbec12

[19] Ellen Nakashima. "Hacks of OPM databases compromised 22.1 million people, federal authorities say". *The Washington Post.* July 9, 2015. https://www.washingtonpost.com/news/federal-eye/wp/2015/07/09/hack-of-security-clearance-system-affected-21-5-million-people-federal-authorities-say/?utm_term=.8fb0fa35dcb3

Index

A

Anonymity, 24, 41, 43, 46
ARPANet, 5
Asymmetric Effects, 46, 61
Attack Surface, 10
Attribution, 24
Audacity, 42

B

Backwards Compatible, 8
Bacon, Sir Francis, 1
Beacons, 27
Bugs, 9, 11, 12, 38
    Fuzzing, 12
Burned, 51

C

Cognitive Bias, 22
Collateral Damage, 51
Cost of Entrance, 50
Cost of Operations, 50
Cyber Conflict, 2
Cyber Kill Chain, 49
Cyberspace, 1, 62

D

Deception, 27
Default Password, 9
Defense, 54
Denial of Service, 55
Department of Defense, 62
Dependencies, 48
Deterrence, 20
Device Drivers, 8

Director of National Intelligence, 63
Distributed Denial of Service, 56

E

Economics of Force, 50
Encryption, 26
Exploit
    Zero-Day, 12
Extensibility, 8

H

Hacker, 1
    Advanced Persistent Threats, 61
    Tradecraft, 61
Hardware, 1
Heterogeneity, 53
Honeypots, 27
Human Struggle, 62

I

Initiative, 44
Intelligence, 63, 64
Intelligence Community, 63, 64
Intelligence Gain-Loss Assessment, 43
Internet of Things, 12

J

Jurisdiction Exploitation, 43

K

Kinetic Attacks, 7

Kinetic Warfare, 2
Kryder's Law, 6

L

Locard's Exchange Principle, 41

M

Malware
    BackTrack, 33
    Stuxnet, 29, 42, 58
    Zeus, 33
Metadata, 26
Metcalf's Law, 6
Moore's Law, 5

N

Nielsen's Law, 6

O

Offense, 54
One-and-Done, 54
Opportunity, 43

P

Patience, 45
Persistance, 45
Phishing, 36
Plug-In, 8
Polymorphism, 49
Principles of War, 2
Privacy, 24
Protocol, 64
    Dynamic Host Configuration Protocol (DHCP), 25
    Network Address Translation (NAT), 24
    Port Address Translation (PAT), 24

R

Reproducibility, 48
Resilience, 52
Retrospective Analysis, 15
Risk, 43

Risk Management, 64
Risk Management Framework, 17
Robert Metcalfe, 48

S

Social Engineering, 37
Social Media, 48
Software, 63
    Antivirus, 17, 61-64
    Applications, 63, 65
    Beacons, 61
    Bugs, 65
    Honeypots, 63
    Malware, 61-65
    Open Source Tools, 32
    Operating System, 30, 63
    Security Control, 15
    Third-Party, 30
    Vulnerabilities, 61, 64. 65

Speed, 46
Strategy
    Active Defense, 61
Surprise, 46
Swatting, 37

T

Threats, 7
Tor, 26
Tragedy of the Commons, 64
Transitive Trust, 28
Trolls, 21

V

Vulnerabilities, 9, 10, 15, 48, 61, 65
    Patching, 62, 64
Vulnerability
    Zero-Day, 30

W

Watering Hole Attack, 37
Weaponization, 34
Whaling, 37
Window of Vulnerability, 34
Worm, 49

About the Authors

Dr. Michael VanPutte began his career as a U.S. Army Airborne Ranger, and led troops in combat in Iraq. He taught computer security and cyber warfare at the U.S. Army War College. Next, he ran offensive and defensive strategic cyber warfare operations, working side-by-side with U.S. cyber experts, law enforcement officers, and intelligence officers at the Joint Task Force – Computer Network Operations, and Joint Task Force – Global Network Operations, the predecessors to U.S. Cyber Command. Later, as a Program Manager at the Defense Advanced Research Projects Agency (DARPA), he created and led national research and development projects. He's currently a co-founder and Chief Scientist at Provatek LLC, which conducts research and development of advanced computer network operations (CNO) and intelligence, surveillance, and reconnaissance (ISR) and industrial control systems (ICS) capabilities.

Dr. VanPutte received a BS from The Ohio State University, an MS in Computer Science from the University of Missouri-Columbia, and a Ph.D. in Computer Science from the Naval Postgraduate School. Find out more at www.mvanputte.com or email him at michael@mvanputte.com.

Thomas Sammel began his career in the Marine Corps as an infantry officer, serving as a rifle company executive officer during Operation Desert Storm. In 1994, then Captain Sammel became a Communications and Information Systems Officer, serving in multiple capacities in the Information Technology and Information Security segments of the Department of Defense. He served with the Joint Task Force Global Network Operations, and has led both the Marine Corps Red Team and the Department of Defense Computer Emergency Response Team.

Major Sammel is also a co-author to the inaugural National Military Strategy for Cyberspace Operations.

Following his retirement for the Marine Corps, Mr. Sammel joined Dell SecureWorks and worked on its Incident Response Team, leading the response efforts to over 150 client breaches to cyber-attacks, including multiple Fortune 500 companies. Following his time at SecureWorks, Mr. Sammel became the Chief Information Security Officer for KKR & Co. Inc. Mr. Sammel is currently the Director of Threat Operations at Provatek, LLC.

Mr. Sammel has a Bachelor of Science degree in Chemistry and a Master of Science Degree in Information Technology.

www.ingramcontent.com/pod-product-compliance
Lightning Source LLC
Chambersburg PA
CBHW022128170526
45157CB00004B/1789